Charles Murray Adamson

Sundry Natural History Scraps

More Especially About Birds

Charles Murray Adamson

Sundry Natural History Scraps
More Especially About Birds

ISBN/EAN: 9783337025601

Printed in Europe, USA, Canada, Australia, Japan

Cover: Foto ©berggeist007 / pixelio.de

More available books at **www.hansebooks.com**

SUNDRY
NATURAL HISTORY SCRAPS

MORE

ESPECIALLY ABOUT BIRDS.

BY

C. M. ADAMSON.

NEWCASTLE-ON-TYNE:
PRINTED BY J. BELL AND CO., RAILWAY BANK, PILGRIM STREET.

1879.

INDEX.

	PAGE.
A Naturalist's View of Close Time, &c.	1 to 18
Wild Fowl in the London Parks and elsewhere	1 ,, 10
Protection of Wild Birds and Preservation of Wild Fowl	1 ,, 13
Migration of Birds	1
Migratory Birds	2
Do.	3 ,, 7
British Birds in India and America	7 ,, 10
Stray Thoughts about Birds	11 ,, 17
The Woodcock	17
Sandgrouse, &c.	18 ,, 27
On what are called Genera and Species	27 ,, 30
On Hybrids, &c. (Grouse, &c.)	31 ,, 36
Observations on Breeding Place, &c., of Knot	36 ,, 41
The Knot and other Shore Birds	41 ,, 49
The Sanderling	49 ,, 52
Spotted Redshank and other Birds	52 ,, 56
Vernal Migration of Little Stint	56
Spoonbills shot in Suffolk	57 ,, 60
Common Sandpiper	60
Schinz Sandpiper	61 ,, 63
On the alterations time, &c., may have made in some kinds of Animals	63 ,, 68
The Golden Plover	68 ,, 70
Wild Geese	70 ,, 72
Wigeon	72 ,, 73
Eider Duck	74
Sphinx Convolvuli	75 ,, 77
The Shearwater, &c.	77 ,, 78
Water-rail's Nest in Northumberland	79
Prestwick Car, Birds frequenting	80 ,, 93
Little Bittern	93 ,, 98

PREFACE.

HAVING for many years taken considerable interest in Birds, more particularly Water Birds, and having collected some of the kinds in their various states of plumage at the different seasons, I have sometimes made notes respecting them. From such these scraps have been compiled, and I venture to hope they may afford amusement, if not some trifling instruction, to those who may have the opportunity to see them.

An apology seems due for my venturing to write on what is perhaps usually considered a thread-bare subject, and especially for putting these scraps into their present form, but the lover of Nature seems never to become tired of reading another's ideas, who has devoted some of his attention to her works; and really, when we consider that the more we know we only find out how truly ignorant we are, it gives one some encouragement to try and find out more about them. Whatever number of us there may be interested in such like pursuits, each one will most certainly observe things differently and represent them according to his ideas; and I would guard all readers against adopting any particular writers' theories or statement of facts or probabilities, so different are men's views, and so will be their manner of representing and describing them.

No one can expect his work to be without many errors, but I need only add, if the perusal of these miscellaneous writings (which I fear contain many repetitions) give any one a slight stimulus in the direction of following a most interesting study my object is attained: my having made notes originally was to acquire information and not to give it.

Much of what a man knows must die with him at any rate, however much he writes. I believe, however, those who do write, on reading afterwards what they have written a few years before, if they wrote again would find that their writings must contain many alterations, and even perhaps contradictions.

Such a subject as the present is inexhaustible, and there will always be food for the mind for all beginners to enquire after and to find.

I do not venture to take the trouble or responsibility of printing these scraps except merely as presents to those I meet with who might take interest in them, and I trust they will not criticize my occasional employment too severely.

It is impossible for me to exaggerate the amusement and pleasure I have derived for years in acquiring the information which is contained in these scraps, and these must be taken as the apology for my conceit in wishing them not to be entirely lost sight of.

North Jesmond, Newcastle-on-Tyne,
  October, 1878.

Perhaps I may add a catalogue of my collection, with the dates of capture and states of plumage of the birds in it, which in itself would show to a certain extent the time of their arrival on our shores and of their departure, and with some remarks on some of them.

A Naturalist's View of the Extension of the Close-Time of the Sea-Birds Preservation Act in Northumberland, and on the protection of wild Birds generally. By CHAS. MURRAY ADAMSON.

A FEW weeks ago I observed in the newspapers that the Home-Secretary had extended the close-time in this county, Northumberland, from the 1st day of August to the 1st day of September.

I presume this has especial reference to the birds breeding at the Farne Islands. These Islands are I believe private property, and are let at an annual rental: the consequence of this is that a great number of the eggs are gathered for sale. In addition to this the poor birds during the summer are disturbed and ill-treated by the numerous steamboat loads of people, often lawless, whom I am told sometimes overawe the few persons residing on the Islands who ought to be the bird protectors. Under the circumstances, however, I rather question their being so. If they are allowed to make as much money as they can from the eggs, they of course take those eggs which are most uncommon, and which bring the highest prices, and thus the rarer species suffer most.

When I visited these Islands early in July, 1875, the keeper of the outer lighthouse, who said he had been there only a short time, told me his predecessor had a large family, and that during the summer time they had lived almost entirely on the eggs. He had himself a number of eggs ready blown, several of which I bought for my children at three shillings a dozen. At that time there were some young birds, among them Puffins nearly full grown, one of which we took from a hole and returned: there were also eggs of most of the kinds of birds which breed on the Islands. Near the end of the month there were young Gulls and Terns, some flying, on the inner Islands, but we could not find one young Oyster Catcher, nor did we see old Oyster Catchers flying about and showing, as they always do by their anxiety, that they had young. I therefore presume none got away that year, for we looked pretty closely, as the commander of the gun-boat whom we were with wanted one for a pet. That year one Sheldrake's nest was found on the Island, but our informant living there told us he had taken the eggs himself.

That year I do not think any young Cormorants got away, as, late in July, this species all had recently laid eggs, many nests having only one egg, and not a young bird was to be seen. Several young Terns were flying, but there were still plenty of Terns' eggs. Probably many of these, in consequence of the lateness of the season, would never be hatched, and if they were, the old birds would very likely forsake the young before they arrived at sufficient age to take care of themselves, as the old birds of some species seem to leave their breeding grounds at a fixed time, probably to get to some other locality in time to moult or for some other reason we do not understand. The consequence of the destruction of the eggs is, that the birds do not hatch their young at the time Nature intended they should, and their season is altogether disarranged.

We saw no young Eider Ducks following the old females, many of which were in flocks. I am quite willing to admit the gross cruelty practised at these Islands by summer visitors, but the worst part is perhaps not the shooting; the young birds are wantonly destroyed, and in 1875 the grass on the Island inhabited by the Puffins was set on fire and many of the old birds, as well as the young, were burnt in their holes. How is a repetition of this to be prevented? In 1874 some young Black-backed Gulls were not able to fly in September, and we saw at that late season a young Eider Duck only recently hatched.

During our visits to the Islands in 1875 we did not see the Roseate Tern; nevertheless it might be there amongst the number of birds and escape our observation, although this bird's cry and flight make it conspicuous to any one knowing it. One of the Outer Farnes was frequented by the Arctic Terns, and one of the Inner Farnes by the Common Tern.

Under the present law it would appear that the owners of land have no more right to take the eggs on it except such as are especially protected by Acts of Parliament, than a stranger; and I have been informed, that some years since the question about the right or legality of taking the Farne Island eggs was put to the test. Some men were summoned for taking them; they employed a lawyer to defend them; and when the case was to have come on it was withdrawn, the prosecuting parties having discovered there was no law to reach the accused. Generally speaking, the kinds of birds which suffer most are such as breed in numbers together. When they breed in such numbers together and the eggs are so easily found, great destruction may be made in a very short **time. It is** not so where the birds are evenly distributed over a large tract of country. The **only remedy** against such wanton mischief would be an Act of Parliament causing eggs and all other produce on lands to belong to the proprietors of the soil. This might be done, perhaps, by a Permissive Bill, whereby the proprietors might be able to protect their property or not as they thought **right.**

Now, I **would not for one moment** defend the destruction or wanton molestation of breeding birds; I would have them strictly **protected**; but it does seem unnecessarily hard to close the time **for shooting birds in a whole** county in order to protect them on **a small** portion of ground where they ought to **be private property.** At the present time it would appear to be **contrary to** law to prevent any person landing on the Farne **Islands between** high and low water mark, and even if they landed above high water mark, it would be difficult to prove damage on such barren ground.

In the latter part of July, some years since, Mr. John Hancock **and I visited, with Mr. Losh who then resided** in the neighbourhood, a breeding place, **in Lancashire** or Westmorland, of the **Lesser Black-backed Gull, the** commonest Farne Island bird; and they, **young and old, had left the** district, excepting one **young bird, which was fully fledged, and an old** bird, probably its **parent. We** then went to **Walney Island** where the birds were protected, the Black-headed Gulls and the Sandwich Terns, the **latter** also a common Farne Island bird, had all flown and gone; whilst at Fowley, where the birds were not protected, **the Common Tern,** and the Roseate Tern, which was there far from un**common, still** had eggs and young. We **were told** that on this **Island the eggs** were gathered to be used in making varnish.*

---

* It was during this expedition Mr. Hancock shot the Tern with a cockle on its beak, **which he mentions in his Catalogue.** It was a rather exceptional bird, I think an imma-

Now from this fact I cannot help thinking, that if the Farne Island birds had fairplay, the young would have nearly all flown by the beginning of August.

A naturalist knows that we have on our coast Whimbrels and Arctic Gulls by the end of July or earlier, whose nearest breeding grounds are the Orkney and Shetland Islands. The Redshanks' young, when not molested, can fly by the middle of July; they and the Dunlins have by that time left the moors and wastes to come to the sea side. The Turnstones have come from Norway by the middle of August, and by the end of the month some of the young birds even from the Arctic regions have arrived. By the middle of August the Spotted-Redshank has come from Lapland, the Wood-Sandpiper and Green-Sandpiper from Sweden; and not only this, but by August nearly all the Curlews and Golden-Plovers have left the moors, and it seems hard if these and such like migratory birds may not be shot on the shores away from their breeding grounds before September. Landowners would hardly like not to be able to shoot Snipes, Plovers, Wild-Ducks, or Teal, should they meet with them when Grouse shooting in August; or because some extensive proprietors chose to take all the first Grouse eggs off their manors to set under hens, or for any other purpose whatever, whereby there were nothing but cheepers on the 12th of August, to have the general shooting postponed for a month.

A general close-time till September would prevent several sorts of birds being procured at all in this country, some migrating entirely before the end of August; and it seems almost superfluous to mention that the sea-birds we have generally in summer are almost entirely a different set from those which are with us in winter. So far as the mere existence of species is concerned, it is doubtful whether the total destruction or driving away of

---

ture Common Tern, a young bird of the previous year, having the front of its head white. From birds in similar plumage, which are only occasionally met with, we may almost suppose they do not breed the first year after being hatched, but that probably the immature birds wander over the ocean and do not congregate with the breeding birds at their breeding stations till they arrive at maturity. The circumstance was singular in this respect; it showed that Terns feed otherwise than by dropping on their prey from the air into the water, a fact which I was not aware of before.

those at the Farne Islands would materially affect them as species, so numerous and so widely distributed are they, any more than the drainage of the Fenlands did the interesting species formerly inhabiting them. Wholesale destruction at one breeding locality drives the birds to other places, but does not extirpate the species.

Under these circumstances, I rather doubt **the** utility of extending the close-time. If the public become aware of the injury they could do by robbing the eggs and young there might be a still worse chance for the birds rearing any young at all. The birds are persecuted in this respect **far too** much already; but if by extending the close-time **you** exasperate the public, they may retaliate **and do more** mischief still.

The shooting **naturalist's** real pleasure in meeting with a rare **bird, or** a bird **in a rare state** of plumage, far exceeds any fancied **pleasure a man can have** in slaughtering birds and unoffending animals by hundreds for any other purpose. All laws should be **as** much for the advantage of the poor as of the rich, and it is hardly right that the rich man should be able to break the law on his own extensive domain with impunity, and yet desire to deprive many a poor man unnecessarily of an almost innocent recreation, and this merely because some persons abuse the privileges so long accorded to Englishmen.

The Act alluded to omits the poor Cormorant from the protected list because Nature directed it to eat fish: rather a curious reason! Probably in Scotland they ascend the rivers and take a few young Salmon, and in consequence they are considered "vermin." Now, if the charge against them is that they feed on fish, why preserve the other fowl at all? They all eat fish, and if the quantity required to provide those at the Farne Islands for a single **day** was taken into consideration, I wonder the fishermen do not take alarm at once, for fear they would clear the sea of all the young fish. Even the fishes themselves, as the Cod and others, not only prey on other species, but even on their own; why are they **not** considered vermin also? Probably the fishermen are trying to clear the **sea of them**; at any rate they take all they can for their own profit, but still the numbers decrease not, except, perhaps, in a few localities.

Nature understands her own arrangements best; she provides food for all her creatures. It seems impossible for a thoughtful man on fair consideration to imagine that the few fishes Cormorants eat could interfere with the Salmon supply. When the young Salmon go from the rivers to the sea, they have to run the gauntlet of many dangers before again entering the rivers. All Nature's creatures must pay their portion of debt; they were all made to eat and to be destroyed or eaten. In Nature's wonderful arrangement there are, with very few exceptions indeed, generally plenty left to continue the species and for all other purposes.

It may I think be taken for granted that there are always more than double the number produced of every species than is necessary for reproducing and multiplying the species; the rest are for the good of other species in the shape of food.

I would hope that the extended close-time may be treated as a dead letter except in cases where punishment may be justly merited. There are several species of shore-birds not mentioned in the Act. Are we to understand the order to be a prohibition to carry a gun at all till the first of September? or who is to decide what may be shot? A person surely should not be summoned for shooting birds not mentioned in the Act; but how is this to be managed, as it is not possible to turn either a magistrate or an informer at once into a naturalist? The only method, I can see, to preserve the birds, would be to make the eggs belong to the owners of the ground, and leave it to them to prosecute for theft or not as they choose.

I cannot help thinking that some method might be adopted for preserving the birds and their eggs on the Farne Islands. These islands are one of the most interesting of the breeding places of sea fowl on our coasts, and if they were in the keeping of Government or the Trinity House they could be easily protected (as it is only in fine weather that boats can land) by the men who are in their employ, who, having good glasses, could distinguish boats at a great distance. Could it not be arranged to grant to some boats license to land, the licensed boat-owners to be answerable for the conduct of the parties they take; and for a rule to be made that no guns were to be fired within a prescribed distance of the Islands during the time the breeding birds are there without special permission?

Since the foregoing remarks were written the Act commencing close time from February 15th has come into operation, and at the risk of appearing presumptuous and wearisome I offer some further remarks on the apparently confused legislation respecting the protection of sea and other birds which seems to me to be entirely in the wrong direction.

The various species of birds, **except** the few wingless species, whose distribution appears always to have been extremely limited, and many of which have become, and others appear to be **becoming, extinct, are nearly all more or** less migratory, either **from one part of the world** to another, or merely from **one district** to another, and often these migratory flights are to a great distance and very extensive. Nature seems to **have** guarded against the possibility of these species becoming extinct by the great extent of country over which they are distributed. Most persons seem to suppose that the comparatively few individuals of each species which visit us represent the whole of the species, instead of their being only a very insignificant portion of it.

As stated above, I **have** almost come to the conclusion that, with few exceptions, it would be immaterial, so far as the existence of the **species is concerned, if** all of those which come to us in **one year were destroyed. I believe** in a remarkably short **time we should be as well supplied as before** with such species **as are suitable to the face of the country** under any altered conditions.

It **would** seem, if we look at the north polar region of a globe, that there is a portion of the world extending probably four hundred miles about the pole uninhabited by birds. Then comes a belt of many hundred miles, partly of land, over a very great portion of which no civilized man has yet traversed. This is the tract of country which produces the hosts of migratory wild fowl which **spread themselves over** the globe southward in winter to get food and to form food for others. The few birds killed from these hordes by man with all his contrivances, nets and decoys included, both in going to and coming from this breeding ground, amount to nothing. The prohibition against taking these birds whenever they can be procured in this country is simply **of no use** so far as extinction of the species is concerned. Each species undoubtedly has a very extensive breeding ground, and the supply from it is and will be **next to inexhaustible** so long as those lands remain in a state of nature. **Neither can you** bring back the birds which have been forced to leave **us; not** that they have **been shot,** but because the character of their former breeding grounds **is** altered and limited to such an extent **as** to prevent them having room to obtain sufficient subsistence. **This cannot** be helped; but what is wanted is protection for the species which **will condescend to** remain with us under the altered circumstances. **It is entirely** out of man's power to make a migratory bird **remain a day beyond its** appointed time, and it always will be so.

There are no **such** things **as rare birds; those which** appear such to us are in reality only birds out of their regular places, and they require no protection. The birds requiring protection are those which either are or ought to be our useful and familiar friends. We should take care of these. The others are not worth protecting; they only afford interesting specimens for a Natural History Collection, which really is their fittest place, as they illustrate the extent to which these erratic species travel.

It is not shooting migratory birds or even snaring and **trapping** them in decoys which materially lessens their numbers; **this is proved by** the extraordinary number of wild fowl **often** taken in **a** single day, and their constant replacement day **after day** so long as the locality suits their habits and the season is favourable. Also the number of Quails taken in Europe, even when journey**ing to** their breeding grounds, which practice has prevailed a length of time, does not materially lessen the supply for the next year. The number of Skylarks netted each winter and sold in the London markets also corroborates this. None of these methods of destruction appear to have thinned their ranks to any appreciable extent; but what drives away the birds from us are chiefly the alterations made by man in the enclosure and drainage of the land.

No one can determine Nature's reasons for having assigned peculiar limits to the breeding grounds of birds, nor can we extend those of migratory species. Man may, and has introduced and acclimatized some species, but these are very limited in number, and perhaps those which he has succeeded in acclimatising or domesticating were especially intended to be made subservient and useful to live in this especial manner.

I do not in the least like sentimental grievances, but any law prohibiting the killing of migratory birds on mud flats and such like unreclaimed places is both unjust and unnecessary. No one can pretend that birds frequenting such places either do or ought to belong to him. One day they are perhaps in Africa or on the shores of the Mediterranean; they are perhaps next observed on the Norfolk coast, and the next place they may be met with is, perhaps, in Lapland, Iceland, or even North America; besides, it is only those who are at the trouble to follow them into their haunts who see them alive. So soon as they have rested long enough, off they go, and the flats are left bare, and no one would follow the few birds left, except perhaps some enthusiast in the hope of meeting with something rare, which, if procured, would do no harm whatever, as it would only be a straggler out of its proper place. Such migratory birds are totally independent of man and all his devices to extirpate them, and against these they can and will hold their own. Should man be able to control the tides, even still he could not prevail against these birds; for if their feeding places were entirely destroyed here, their powers of flight would enable them to pass on unheedingly to another place; but so long as man cannot control the tides, there will remain plenty of resting places for them, where they are often unapproachable even with punts or other contrivances; and if they are approached and a few are shot, the remainder seldom alight again within any reasonable distance, and cannot be followed. It is totally different with such birds as Partridges and other game birds which are local in their movements, and at times may be followed up and the whole covey obtained on a limited **tract of land.**

Many **persons believe** they can treat the **Woodcock** like the Pheasant, and suppose he will alter his character **by** protection during the breeding season. He is a wanderer by **nature, and you cannot clip his wings** in order to make him stay, **and no protection** here in summer will make any appreciable addition to his numbers in winter. **Far be** it for any one to imagine **I wish him** interfered with during **breeding time,** but as he generally **keeps to** woods, let the **owners look after** him when there, and do not enact new laws **for his safety.**

Nothing having life except **man can be cruel; he alone is** endowed **with mind, and is capable of knowing right from** wrong. Lions, Tigers, Hawks, and Dragon-flies were all **created to be selfish, and in all** their depredations they are only carrying **out their Creator's intentions,** and we have no right to call them **cruel. Our doing so is setting** aside the Creator's wisdom in having made them. **I contend that any one killing** the apparently most innocent and **inoffensive creature** for inspection **or** admiration is not cruel; **in fact quite the reverse, for** by so doing he is honouring his Creator by admiring and wondering at the operation of His **power. Cruelty consists in** the wanton and thoughtless killing, or sanctioning **the killing,** unnecessarily, **of any** kinds **of** living things, all of **which, even** the most repulsive, **were created for some good** purpose, which perhaps man, with all his intellect, cannot at once perceive. **There are many who would be very severe** in condemning another for **killing a rare bird or butterfly, but** who daily sanction the killing of a **Hawk,** when there is really this difference **only, the** one is probably **killed** for a **good purpose,** and the person who killed it has **some** thing to look at for the remainder of his life with pleasure, **whilst** the other is killed for mere caprice or fashion, as in all probability the day after it would have been far away from the district **never to return, and therefore the** injury it could possibly inflict **was only imaginary.** Perhaps, too, the treasure the former had secured might have been destroyed immediately after in the ordinary course of nature, so that it mattered little to it, bird or insect, whether it was taken or shot by man, or had been caught and eaten by a Hawk or other bird, its natural end, for Nature seems to have ordained that few, if any, of her creatures, birds, beasts, fishes, or insects, should die what we call a natural death. All, as soon as they get feeble, **and very often before**, merely **form food for others.**

The **only legislation** really required is protection **for all our summer** residents during breeding time, including eggs **and young of all** breeding birds, great and small, at this season, **on** the grounds of humanity alone, unless they become a positive nuisance.

The war of extermination carried on during breeding time against Hawks, Owls, Crows, Magpies, Jays, Wood-Pigeons, **and others whose real habits we really** know little of, is at **the least cruel and unjustifiable when we take** into consideration the **numbers of young birds left to** die in the nests **when the old birds are** trapped. Morally there is no greater **harm in** any one torturing a domestic animal than there is in starving a wild one. I would not think it right to prevent a person **shooting any of** these birds he chose on his own land, **but I hope the time will come** when we may not all **be so addicted to fashion, and that** some large landed proprietors will take delight **in having on their estates all kinds of birds Nature intended to be there,** and in observing **their habits and hearing their cries, all of which add charm to a** country life. **One of the greatest drawbacks to the country is the absence of the** various birds which **ought to be there; and which of us will or can take** upon himself to say that **the good such birds as the Magpie** and Jay do in destroying insects **and their grubs during the** whole year, does not quite compensate for any **injury** or imaginary injury **they do or are** supposed to do during their breeding season.

I would ask whether Rooks do not take eggs when they can **find them?** Birds are generally cunning enough to conceal their eggs, **but** they are probably as likely **to be found by** a Rook as a Magpie. Why then make enemies of the one kind and spare **the other?** The chief reason seems to be custom only.

As matters are at present, persons living near villages and towns are continually annoyed by having their gardens entered entirely against their wish, **by** boys and **others** seeking birds' nests, but there seems to be no legal remedy for this. The only way to protect birds successfully would be to make **them, their** eggs, and young, belong to the person or his tenant on **whose** ground they are, and to enable such owner or occupier to have any one who takes either the one or the other punished, at any rate if he chooses. **Thus all trespassers** looking for any kind of eggs, and whatever comes in their **way, might be** stopped; and we might really do without **game laws altogether**; and be so **much** the better in this respect, **that each person** might then do **what** he liked with what ought in my opinion **to belong to him, but** which in reality does not legally.

All legislation should be on the plan that what **is morally right** should be legally right, and *vice versâ*.

The laws respecting the time when our resident game may be **sold** and taken do not seem **to** be very hard, as no one who wants a bird **is** deprived from **getting it during the greater** portion **of** the year, but still there is **not so great** a difference now-a-days between the value of a Partridge and a **Golden Plover, and the** eggs of each should equally belong to **the persons on whose** grounds **they are.** There certainly is this **difference, that the** Partridge remains where bred or near **to it, the Plover, as soon as he can fly,** will change **his ground, but so long as they have their nests** on a man's ground they should be his **property.**

As I said before you can do no harm by shooting birds so long **as they remain in flocks, and it** seems very hard and quite **unnecessary to legislate that birds,** whose breeding places are **confined to the extreme north, such species as the Grey Plover, Common** Godwit, Knot, **Sanderling, Pigmy Curlew, Little Stint, Temminck's** Sandpiper, **Spotted Redshank, Turnstone (Greenshank and** Whimbrel rarely breeding **in Scotland), and others which visit us** on their migration **northward so late as April and even** May may not be shot **by the naturalist or sea-side gunner, in** order to secure specimens **in the interesting plumage many of** them put on at this season. **No legislation will make these**

birds **linger** a single day, and as they never did breed in England, even in its wildest state, they certainly never will now.

Many species of birds, the Spotted Redshank, Common Godwit, Little Stint, Temminck's Sandpiper, and Pigmy Curlew come to our own coast, almost exclusively in the young state, immediately after being able to fly, and after staying a very short time, perhaps a few days only, leave us never to return. What harm arises from shooting these? It is quite clear *they* do not **come back.** The fact is, that for many of our migratory birds, Geese, Ducks, Sandpipers, and others, the immense morasses in the North of Europe are their summer **home;** thither they wend their way, arriving late in May perhaps, and taking only a week **to** get there, as soon as **the grounds are** fit for them after the **long winter, and then of what** consequence is it how **many we can capture at this season? Each** species differs in **some respects; but after the young of** several species are sufficiently strong, in about eight or nine weeks, off they come **back** again scattering themselves southwards over the face of the earth, not only to find food for themselves but to form food for other creatures, and also to afford instruction and amusement to man in studying their habits. Those who are not **above** doing this are not always unprofitably employed, **and** any legislation preventing **this** is **wrong.** We find this habit also to be common to our summer **residents which depart** with their young, hatched **and reared here, but** which return the following spring **in** about the average number only of previous years, all the rest having most probably formed food for other species during their absence. For this purpose they were made in accordance with one **of** Nature's laws, and it teaches us that what may be called **over** protection or care for them often avails nothing.

I mentioned before that so long as most birds continue **in flocks** little **harm** arises from shooting them, and I think this is **so.** When in flocks they are generally not near their breeding grounds: when they come to these, they generally do so in a quiet and **un**obtrusive manner, and spread themselves widely in pairs.

Birds coming from southern climes to breed with us, take for instance Shovellers, generally arrive at a pond about the 20th March. They know that by that time there will be no lasting frost to prevent their getting food, but when they do come, they come so quietly that you find them at the pond and there only, none being observed perhaps for miles round about, either in coming or departing after having reared their young.

During the present close-season Garganies would never be killed in England, as they only arrive about the 20th March when they do come, and depart as soon as ever the young can fly, August at the latest.

Those extensive landed proprietors who have honest servants, and who possess large tracts of waste land with shallow water which stands long enough to produce the food the birds require, are independent of all laws. So long as the marshes are extensive enough to enable the birds to take long flights on being disturbed and to get into a place of security, then they will congregate and remain in them, but only so long as Nature has taught them. If often disturbed, or if the tract of ground is not sufficiently large to guarantee them security from frequent molestation, they will not remain.

This district is so much what is called improved, that there is now hardly room for real wild fowl anywhere. Even when there are ponds, the country round about is not wild enough and extensive enough for them to remain at.

Many persons have what they call wild Ducks on their ponds. These have been placed there by themselves or their predecessors. They are very different from the truly wild birds in their nature, seldom going far from home. The truly wild birds will not be controlled in their habits. They prefer being free as the wind, and as changeable. They pair in spring and keep themselves aloof from the half-castes, very seldom indeed having any variation in their colour. The others are polygamous and often change their colours, having always coarser feet and other parts. They are no doubt valuable as ornaments, as well as for food, when you cannot have the true wild birds. Nature does not countenance variations in colour generally in wild birds and animals, but there is no rule without an exception, instance the Ruff in summer. Almost all her productions are similar according to the seasons, ages, and sexes, but why some species vary more than others according to age, sex, or season, is perhaps what we can never find out. All Nature's species were created perfect, and if left to themselves remain so and constant. The several **varieties** produced **by man's** intervention, and such as he **takes pride in** producing **and** keeping **up** are not so; if they were **left** to themselves they would soon cease to exist as such.

There are large tracts of land in England now which were formerly and are still inhabited by birds in summer. The recent acts perhaps prevent the destruction of the old birds, but if the eggs are constantly gathered **you may** almost as well be without the **old birds.** They, **in** consequence of their eggs being **taken, after a time assemble in** flocks, leave the places, **and** are **as wild as in winter; whilst if the** eggs and young were protected, **any one would have the** pleasure of seeing the old birds **near** when attending to their young.

It is perhaps little use trying to alter this state of things. Civilization and population increase so rapidly that the wild birds must give way, but it is a pity not to help them as much as possible, and this should be done by **the** owner of the land, who should legally have the power.

All laws should be made to inflict as little injury as possible on individuals, and they should not be **made** at all unless for the public **good in** some way **or** other. I cannot see how any possible good arises by preventing birds being shot during their migrations, **and which are here to day and** probably hundreds of miles away **to-morrow.**

Real injury accrues to individuals **when resident** birds are shot, or their nests taken in private grounds. Most persons who **have any** kindly feeling generally take interest in having birds **and** their nests in their grounds, and having them often destroyed **is** as great an injury to them as it would be in having something generally thought much more valuable taken away or stolen.

It would I think be better not to have a close time at all for mud flats and such like places. No birds breed on these, and as soon as **the** birds disperse to breed, then such places are almost entirely forsaken by them. Let the landed proprietors look after their own grounds, give them the birds and **eggs** on them, but make laws so as to enable them to punish trespassers and robbers, at any rate if they choose, and if they do, the birds will **increase.** It is no use protecting such birds as have **ceased to** breed in England from altered conditions of the face of the country, such as drainage of fens, and certainly none whatever protecting such birds **as never** did breed in this country. No protection will make them do this, and all the protection you can give is not even likely to increase **their numbers here.**

All naturalists must regret the circumstances **which have rendered it** impossible for such species as the **Spoonbill, Avocet, Black Tern,** Black-tailed Godwit, Bustard, etc., **to remain and breed with us;** but these altered circumstances **were for the good of** the country generally, and the consequent loss cannot be helped. How much more should we regret the extinction of **the** Kite, Moor Buzzard, Common Buzzard, Hen Harrier, and some other equally ornamental species, **which have nearly** become extinct from man's personal greed : birds, too, which frequently were satisfied with food which was entirely useless to man.

There is one thing I think which good taste should immediately put a **stop to, and that is, the** extensive slaughter of game **by persons in high position, so often** reported in newspapers. It is said imitation is the sincerest flattery. If an ignorant **man reads that some great man has** killed several hundred head of game or Pigeons, if he thinks at all, he must see that it was done merely for the love of slaughter and not for love of manly **sport.** Well, he of course has not the game **to shoot at,** nor **can he pay for the** Pigeons, but instead of these **he gets** amongst some unfortunate young Gulls or Terns, **and** he goes on slaughtering them till **he has, as far as** he can, matched his beau ideal of a shooter; and how **can** we blame the one without condemning the other? **If** persons having the means and the desire to **kill** such numbers of birds or beasts do it, it would be well if **they kept their** exploits **to** themselves. Besides this, by informing **the public they invite** poachers to try their preserves, and thus often get their keepers into trouble.

Taking the whole system of game-preserving into consideration and the expenses attending it, it would seem almost a question, now that money is taken into account in all such transactions, whether the great expense required in preserving is compensated by the equivalent amusement and money returns for the game got and sold. That is, it seems a question if what are considered vermin were allowed to exist, and trusty servants only kept to protect their master's property, birds, eggs, and other things, whether the great saving in expense would not be met by having a sufficiency of game, and a great deal more pleasure in obtaining it, than by the present system of heavy bags, got at enormous expense, and in such quantities at a time, as to be almost useless to the proprietors except in the return he gets in money from the dealers.

I am quite unable to imagine how any liberal-minded naturalists could sanction such changes as have been made by recent bird legislation. Even supposing that they had individually found out all they cared to know concerning the habits of British birds, still they had no right to stifle future investigation by their juniors either in age or learning. Some kinds of birds require different times to arrive at maturity; several species of Gulls take four years. I have not yet met with any one who can determine the age of some individuals of these, met with during immaturity. Now we may take it for granted that there are quite as many immature as mature Gulls in existence. The latter only congregate at their breeding grounds, where the owners of such ought to be able to protect them. The former are probably scattered over the coasts and seas in many parts. Now, what harm accrues to these species by killing for observation these immature wanderers? I would ask, do these naturalists think that all information required by a student of nature can be acquired from either stuffed birds or skins already obtained? If not, how is it to be got? Perhaps by looking at the birds with binoculars. If they have to get their information from skins already obtained, I fear they will be as likely to get it as a surgeon would be who expected to learn anatomy from a mummy. A true naturalist requires the recently killed birds to judge of their shape, and see the colour of their eyes, legs, beak, and feathers, which in so many species change during the growth of the bird and with the seasons. He wishes to admire and contemplate their wonderfully different forms, and the arrangements of the layers of feathers, as adapted to the bird's requirements.

I would ask, can any one yet satisfactorily explain the change of plumage, from winter to summer, of so many of our shore birds, which takes place about April? even of the Dunlin, comparatively speaking a common bird? and how much of this change is acquired by moulting? and how much by the change of the colour in the feathers themselves? or can any one explain the length of time birds of these genera take to change from white to black or from white to red? or the cause of such changes? or can they account for the change in the plumage of the males of the surface feeding Ducks, which takes place about July, when they acquire a plumage somewhat resembling the females? and then again the change to their full plumage about October?

The Wigeon and Wild Geese which remain longest on our coasts in spring are most probably not those which would remain in Scotland to breed. The birds breeding there have most likely nests before the flocks leave our more southern shores, where they are lingering till the distant northern morasses are ready for their economy, which would not be till about the middle of June. It appears to me, therefore, to be quite unnecessary and useless to prevent these coast birds being shot, so long as they remain with us in flocks, or at any rate, until the end of March.

It remains for me to ask forbearance from my listeners in their criticism of what they have heard. I have written down my thoughts as they occurred to me, and am afraid I have used a great many repetitions, but they are used as explanatory arguments, and in consequence of my not being accustomed much to writing I have used a crude style, which I hope will be overlooked.

In conclusion, I feel uncertain whether I have thrown any light upon a subject which apparently is very little understood by our legislators.

WILDFOWL IN THE LONDON PARKS AND ELSEWHERE.

(Reprinted from " The Field" Newspaper, April, **1865.***)*

In *The Field* of Jan. 28, H. H. mentions several sorts of wild ducks, with perfect wings, as having been observed by him on the ornamental waters in the London Parks. Some years since I saw several mature Wigeon which could and did fly, and I was as much surprised and puzzled as he is. This bird will not apparently reproduce its species in confinement,* and in a wild state it is a wary bird, and seldom admits of a near approach, even **when pressed for food by severe weather;** its food seems **to consist entirely of vegetable** substances, and it is pretty **to see them** nibbling the grass like small geese. It probably never remained **to** breed in England; it arrives in September, **the old** drakes then having the white patch on the wing, a sign of maturity, but having the freckled portions of their plumage mixed more or less with the bright rusty-coloured feathers **of the summer** plumage yet remaining. Young drakes of the former year sometimes remain till May, by which time they have nearly acquired their full plumage; but many of them have not then got the white patch on the wing. Many young drakes are scarcely distinguishable from ducks till far into winter, only traces of the freckled plumage being visible when closely looked for. This bird seems equally at home in salt or fresh **water,** and consequently remains with us in considerable numbers during the most severe weather; when severe frost sets in it flies direct from the fresh water to the sea, not remaining, as the mallard often does, in sheltered places, amongst bushes in burns and small rivers. Those which

* I put it thus, as I have seen hybrids between the Wigeon and Mallard bred in confinement, also between the Pintail and Mallard, and a hybrid between the Wigeon and Pintail, a drake of which had been sent to the London market amongst other wildfowl.

remain seem to prefer pieces of water with sloping edges and clear of bushes and high banks; it is not often seen in small ponds, but is numerous where there are extensive sheets of shallow water, with grassy edges. Nothing is prettier to see than a flock of hundreds of them flying and wheeling about high in the air, and it is pleasant to hear the rushing noise made by their wings as they descend from a higher elevation to a lower one, as they often do.

The same year I saw the Wigeon in the parks, one of the keepers in the Regents-park Gardens told me the Tufted Ducks had bred, and that they had not been able to catch the young (and no wonder, as, if they once get to the water, they dive like water-rats). Such a circumstance might account for their having been seen flying from one piece of water to another, and might also account for the few instances which have been noticed within the last few years of this bird breeding in this country. That some wild Tufted Ducks and pochards should sometimes find out pieces of water like those in the parks, and, being undisturbed, remain for several days, is probable enough. Both these species often reluctantly leave the water, and will sometimes suffer an approach within one hundred yards without concern. I am not aware that the Pochard has bred in confinement; and, even if it did, probably the young would not remain with the parents unless pinioned, but would follow their natural instinct, and migrate at the regular season.

I never remember seeing a Pochard in its first plumage: all, on arriving here, show most of the plumage of the old birds; and it seems probable that the few Pochards and Tufted Ducks which may have naturally bred in this country, both old and young, migrate early, as it seldom happens that either species is met with in early autumn, or until the ordinary autumnal, or rather winter, migration, when those which have been bred at their regular breeding places appear at particular lakes and ponds in considerable numbers, a few remaining during the winter. Many sorts of wild birds, in severe weather, quickly perceive where they may remain in comparative security. This year I saw Tufted

Ducks and Pochards which sat upon the ice where swans had kept the water open, till you advanced within fifteen yards of them, before they would rise; some of them, even at that distance, not withdrawing their beaks from under their back feathers, and when put up, merely flying round a few times and settling again.

The appearance of the Shoveller and Gadwall with perfect wings is much more curious. The former seems to be a shy bird, almost constantly frequenting shallow, still, fresh watery places, where there are sedges and reeds; it appears to be a delicate bird, and looks unhappy in confinement, particularly in winter. It does not appear to come to its breeding places till all chance of frost is gone, or whilst there is danger of its being prevented thereby from getting food. It is an insect feeder, and often swins about in circles as if to make eddies, and apparently sifting the insects out of the water with its singular beak, keeping its head low whilst thus engaged. The latter part of the month of March appears to be the time for its arriving. I have seen a dozen together by the first week in April. They then fly about a good deal, and appear to be pairing, two or three drakes sometimes singling out one duck and chasing her, and making a singular noise. By the 20th they seem to be paired, and they have eggs early in May. Though the Shoveller is occasionally met with during winter, it should be considered a summer visitor, and no doubt was very much commoner in England before the places suitable for its accommodation were drained. By the time the young are hatched the weeds, amongst which they keep, have become so rank that is scarcely possible to see them; but it seems evident that the birds suddenly take their departure soon after they can fly well, as few indeed are to be met with between breeding time and November, when a chance bird sometimes turns up, and though the young apparently get away, no greater number of old birds return the following year, probably on account of the now limited ground suitable for them.

The Gadwall, in the north of England, is a very rare bird, and probably never was anything else; and had it not been for the quantity of wildfowl sent, within the last thirty years, from the

Continent, it is questionable whether it would not have been a bird rarely seen in collections in England at this day. I have a duck Gadwall which I shot at Prestwick Carr on the evening of Nov. 28, 1850, and I believe not more than one to two instances are known of its having been met with in this county. The evening before there had been a sharp frost, which had covered the water with a coating of ice; quantities of ducks, during the day, kept in the middle, where some water was open, and sat about the edges of the ice. In the evening, two others and myself, one of whom had the adjoining shooting, watched the ducks coming out. I was in the middle, and the first birds that flew near were two which I thought were Mallards. I shot one, and hit the other, which, however, unfortunately went on. The one which fell dead made a hole in the ice, and disappeared under it; and though one of the party, who had as good a retriever as could be, came with it, nothing was visible which the dog could see, and we left the bird, he saying he would get it out at daylight and send it. We again took our places, and I was fortunate enough afterwards to get four Wigeon. The bird was duly sent, with a letter; the writer—being well acquainted with the various sorts of wildfowl he had been accustomed to see—remarked that the bird sent was a curious one, and suggested its being a cross between the Wigeon and the common Wild Duck—the grey wing and white belly resembling the Wigeon, the rest of the plumage resembling the Wild Duck, and the clay-coloured feet being supposed to be between the two. Now how easily this bird might have been set down as a hybrid; and, had it not fallen into the hands of one interested in natural history, it most probably would have been. Gadwalls are most commonly to be seen in Leadenhall market in April. I have occasionally seen them in November, but not in the depth of winter.

The Teal, like the Shoveller and Pintail, dislikes frost and snow, and appears to be a much more inland bird than the Wigeon; it delights, after rain in spring, in the evenings, to walk over the grass land in search of worms; it also appears to be a summer rather than a winter visitor. Numbers come in

March during their migration, many more than remain to breed, though in suitable places it breeds not uncommonly. During very severe weather, instead of resorting to the sea, it seems to quit the country in preference; but, during its autumnal migration in September, flocks are not uncommonly seen at the seaside, composed entirely of young birds. I do not remember seeing an old drake, changing from his duck-like plumage of summer to its full plumage, killed in the north of England in autumn, and therefore ground my supposition that even this species migrates more than is generally supposed; and I question whether those which breed with us do not leave the country soon afterwards, and whether those which come in winter have not migrated from some other country. In spring, in the evening, the approach of a flock of Teal is often announced by the short whistle of a drake. In the Mallard the duck is far more noisy than the drake, whether whilst resting or flying, the more subdued note of the drake being comparatively seldom heard, whilst in the Wigeon and Teal the notes of the drakes are oftener heard than those of the ducks.

The Garganey appears to be the least able to bear cold of the ducks which visit us; it has long been designated the summer Teal, and probably by no chance ever remains during winter. Like the Shoveller, it does not arrive until after all chance of its being frozen out has passed, and, like it, when it once arrives at a pond or marsh during its stay, it does not appear to leave it. I find I first saw three birds, two drakes and a duck, on the 28th March, 1858. On that day I could not make out what they were with the aid of a glass; it was a bright day, but windy, and the water was rough; I was between them and the sun, and the white line over the drakes' heads was very conspicuous. They kept in the shallow water, and were continually turning up to reach the bottom with their beaks. On the 4th April I saw the same three birds, and ascertained what they were as the water was smooth; and the following day I took my friend, Mr. John Hancock, to see them. They remained till the 18th, after which day I could not see them. It is a rare bird in the north of England, and I

B

was surprised at its arrival here so early. On another occasion only I have seen this species alive. On the 13th June, 1855, I saw two, drake and duck, at Prestwick Carr, and after having told a person there what they were, I went again on the 15th to try and get one of them. I found they had been shot, but I saw them dead; the drake had begun to acquire his duck-like plumage; it is possible they had intended to breed there. Now I am almost sure the former three birds were not molested during their stay; neither were a Ruff and Reeve in full plumage, and a Wood-sandpiper, which Mr. Hancock and I saw on the 10th May, 1857, at the same place; and as none of these birds, or others of the same sorts, have been observed since, I think it rather shows that the sparing of rare stragglers, which apparently come to us by accident, and out of their regular course, does not cause any greater number to come in future years: at the same time, I would never think of killing them except for some good purpose.

The Pintail appears occasionally during early winter, but is not a resident; those then got are invariably young, hatched the previous summer, and in immature plumage. A few pass in March, when they are in full plumage, but seem to remain a short time only. Both the Pintails and Shovellers which are found here in winter appear to be chance birds, left behind at the general migrations.

The Mallard, which is unquestionably the most useful, and perhaps also the most beautiful of the whole lot, appears to be and to have always been the most common species; and as it is a general feeder—devouring grain, worms, insects, acorns, grass, and clover—so long as gentlemen keep extensive pieces of water within flying distance, so that when a flock is disturbed at one place it can get rest at another, it will continue to be numerous, in spite of the drainage of the shallow waters and all other improvements. No doubt its breeding places are materially curtailed, but still the enormous quantities of them which come from a distance towards the month of November will always keep them plentiful, in spite of the price upon their heads, and all guns both great and small. The young Mallards appear to have acquired their full plumage before arriving in this country. Though often found at the sea-side, even in open weather, they like to come to the fresh waters in the evenings; and this propensity is well-known to the seaside gunners, who watch at nights for their coming to the fresh-water springs, where they run into the sea. They seem to fly in winter every evening with great regularity as to time, but irregularly as to the course of flight: sometimes they will go to a particular field at a considerable distance from where they have rested during the day, and as they arrive after dusk and depart again before daylight, many places are visited by them without its being known. A very favourable place seems to be a barley stubble which has remained a long time unploughed, and especially if a considerable portion of the grain has been shaken out. Many gentlemen have a sort of half-wild breed of Mallards, which remain in the neighbourhood during the whole year. They are generally—though perhaps difficult to distinguish from their wild relations on the wing, when the colour remains true—pretty easily known when dead: the feet are coarser, and the birds themselves clumsier: they do not show the clean pure breed of the real wild birds, whose condition is adapted by nature to meet all emergencies; and when they have been accustomed to cater for themselves, and rove where they please over land and water, they will not at man's bidding become tame all at once. Notwithstanding the Mallard in his wild state is so much more wary and difficult to approach in the daytime than many other sorts of wildfowl, it is surprising how tame some of these semi-domesticated birds do become. It is probable our ancestors long since tried to domesticate many other species, but found it impracticable to make them useful, and abandoned the experiment. In the Mallard you find the wings, when closed, reach to nearly the end of the tail. If you look at a large tame Duck you will find the body much longer, or the wings much shorter, in proportion, and the consequence is, the tame bird cannot fly: this appears to show that, it being unnecessary for the tame bird and its intermediate ancestors to use the wings to procure food through domestication, the proportions of the structure of the bird have in time become altered.

The White-eyed Pochard has never been met with in the north of England, to my knowledge; and I believe we are generally indebted to the Continent for almost all the specimens we see of this bird in collections. It was not uncommonly to be seen in Leadenhall-market in spring, though rarely in autumn; all I saw were, I believe, foreigners.

The Scaup Duck seems to be a hardy bird, and quite able to take care of itself in the severest weather; it prefers salt water, and its food seems often to be tellinas, which it swallows whole, the broken shells in the gizzard acting as sand to digest its food; in severe weather I have found even periwinkles (*Turbo littoralis*) of considerable size in its gizzard.

All sorts of Ducks seem to prefer still water to rivers and burns, and it is generally only when in winter the former gets frozen over that they are compelled by necessity to take the latter. So many, however, of those species that cannot last out during a protracted storm depart at the approach of severe weather, that whatever happens to those which remain is of small consequence: sufficient for Nature's purposes will have wintered in safety, and so soon as their presence is required at their breeding places they will be there.

The migrations and partial migrations of birds seem intended as a provision against the possibility of the species being exterminated or numerically seriously damaged by severe weather or other accident in one locality. However wildfowl are distressed by and persecuted during a storm, no sooner does it go than they again appear in as great numbers as before, provided the food is there for them. Though bad weather makes wild fowl comeatable, and consequently more are killed by being driven from the places they usually resort to, a greater variety of species is observable during the times they are moving to and from their breeding places. I have seen, during the month of March, the following species on one piece of water the same day, which far outnumber what could have been seen any month during winter—Shoveller, Garganey, Teal, Wigeon, Tufted Ducks, Pochards, Mallards, Pintail, and Golden-Eye; and on a bright mild day at this season, an hour may be pleasantly spent in watching them with a glass, as some of them play about and lash the water with their wings whilst bathing and diving.

In addition to the species I have mentioned, we have on the coast—Eiders, Sheldrakes, Scoters, Velvet Ducks, and Long-tailed Ducks; and I will only say that as little is known of their migrations as of the others. The Long-tailed Duck is peculiar in this respect, that, whereas in all the other species the Drakes, after once acquiring their full plumage retain it till after the breeding season, it does not; but about March his white head becomes black, and the elongated scapulars, which are white in winter, become black and brown; some of them remain into April, and complete this plumage before departing to their breeding grounds.

Many different species of ducks occasionally are locally indiscriminately called by the same name. I have known Tufted Ducks termed Golden-Eyes, and even once heard Mergansers called "Teal." Too much reliance should not be placed on the notice of the observance merely of a bird said to be of a rare species. I know how very difficult it often is, even with the aid of a good glass, to make out, under favourable circumstances, the difference between some species, particularly when not in the distinguishing garb of the male in full plumage. It requires a person to be a good naturalist, as well as an observing man, to distinguish at a distance, in autumn, such birds as Teal from Garganeys, Wigeon from Pintails, and a different sort of a naturalist entirely from him who can tell you what a dead bird is when he has it to examine; and when it is considered how few men really know the young birds and the different states of plumage, it seems probable some species have been admitted as English on insufficient authority—such as the Buffell-Headed Duck, the young of which much resembles a young small Golden-Eye. The drakes of this latter bird do not appear to get their plumage the spring after they are hatched, as, like in the Mergansers, we frequently see drakes little changed to their full plumage far into the spring. Probably, the reason why the mature drakes are so seldom seen in comparison is, that you have the old duck, and perhaps from six to eight young in the immature plumage, which reduces the chance of an old drake being met with. The difference in size between the old drake and a young duck is puzzling, in consequence of her being so much smaller than might be expected.

No sooner does the cold weather disappear and spring come, than ducks commence to moult, and before the end of March they become thickly clothed with down under their feathers. This is for the purpose of enabling them to supply their nests. The old drakes do not commence to moult till much later in the season.

I feel quite certain that very little is known about the migration of any single species, and should these remarks be the means of calling the attention of any one to the subject, I am also certain, though it may be considered of little use, he will find its pursuit amusing, and certainly not disagreeable.   C. M. A.

PROTECTION OF WILD BIRDS.

(Reprinted from " The Field," April, 1877.)

Sir,—What are we doing, naturalists and all? are we not straining matters too far? Some naturalists, who are authors, seem to forget that they must have derived much of their knowledge and pleasure by examining birds recently killed, either by themselves or others, in the various states of plumage several of the species get in the different seasons. I would ask, are they entirely satisfied with the knowledge they have acquired and committed to writing; and do they **wish to** deprive the rising generation from having **similar** opportunities to those they themselves had, and to leave them **to limit** their observations **to what** has been written only? I would fain hope not.

Another question I would ask is, what good we do in protecting birds which are not, so far as we can judge, useful in any way to man, except to study their habits? A number **of the** species we are apparently solicitous about, unless shot, nine hundred and ninety-nine out of every thousand of us would never even see, and those who do see them would only do so after expressly going to look for them at limited places, as well as seasons.

We are hearing remarks **about birds having** been seen. Now, I have little faith in such generally, more especially when particular states of plumage are described, unless they are very conspicuous.

When we **come to** some of the Sandpipers, the difference between the old Dunlin in summer and the just fully-feathered young birds, or the young Little Stint and the summer plumage, how few know the difference when they have the dead bird in their hands; and, as to Buzzards, have naturalists come to any definite understanding whether the pale and dark birds are varieties, or the difference is caused by age?

I would ask, why protect those species which pass us in their periodical migratory flights at all? None of the following waders ever bred in England, or ever will—Whimbrel, Spotted Redshank, Greenshank, Common Godwit, Grey Plover, Knot, Purple Sandpiper, Little Stint, Pigmy Curlew, Temminck's Sandpiper, Green Sandpiper, Sanderling, Turnstone, Solitary Snipe, Jack Snipe. All these are common as species, and very widely distributed; and I will fearlessly, without hesitation, say that any quantity that may be killed here of these at any season will not lessen their number, nor will protection increase it. The numbers of each appearing annually vary from other circumstances, such as prevailing winds and the weather. Many of these species only keep on the sea-shore, where they are often quite unapproachable. By making a close time for these, you prevent some of the species being ever procured; and under these circumstances, really of what use are they? We might as well be without them altogether. There are many things about birds to be found out yet, and I think we can scarcely afford to be without the means of obtaining further information, which we can only get from the examination of recently-killed birds at all seasons. Which of us can tell why the Dunlin gets a black breast in summer, and the nearly allied Little Stint does not? why the Knot gets a red breast, and the Sanderling remains with a white one? why the Grey Phalarope gets red nearly all over in summer, and the Red-necked Phalarope remains with a white breast? At present the most learned of us cannot explain these matters.

So far as the various sorts of ducks are concerned, those which sometimes remain at particular places to breed, such as the Shoveller, Garganey, Pochard, and Tufted Duck will continue to do so, if preserved by the owners of such places; and it may be remarked that, when once they arrive where they intend to breed, they seldom ever stray far away, and as almost every one knows, for several weeks in summer the old birds are totally unable to fly; but even these species are wanderers, and we have no means of knowing whether the few, comparatively speaking, bred at such places, are not the very first to migrate southwards, so soon as ever they can fly well. Certainly the Garganey does, as no birds are ever procured after very early autumn; and I think the

Shovellers bred here leave so soon as able to fly well—certainly the first severe frost scatters them all.

I trust no great body of Englishmen in these enlightened times will try to prevent foreigners sending the produce of their country here, particularly such as birds coming to them during their periodical migration; such interference would appear to me unjustifiable. The ducks are only taken by them so long as they are in flocks; so soon as they pair, and the flocks get broken up, they cease to take them. Those taken are only a small portion of the flocks passing to the north, which do not scatter and pair before leaving the country, thousands never coming near the decoys at all. I do not think real wild ducks pair till near breeding time, as you so often see flocks of Wigeon and Teal composed almost entirely of one sex late in spring.

C. M. ADAMSON.

[Our correspondent seems to have overlooked the fact that all the wading birds, both resident and migratory, may be shot at any time between July and February, when they are not only more numerous, but in much better condition, than they are in the spring. He is mistaken, too, in supposing that some of the species he names do not breed here. See an article on British Wildfowl in our naturalist columns.—ED.]

PRESERVATION OF WILDFOWL.

From the "Field," May 5th, 1877.

SIR,—I have to thank you for inserting my last letter. I feel I am on the unpopular side of the question, but the right one. At the same time I yield to no one in my desire to protect all birds during the breeding season in moderation. What is wanted is to put the matter on a right footing, giving no particular favour to one class more than another.

In your footnote you call my attention to the list of birds I say never breed in England. I omitted Scotland, as I am aware the Greenshank and Whimbrel breed there; but as to the others, it will require more than I have yet seen to convince me I am wrong. Any number of probabilities do not establish a fact. The short time the Green Sandpiper* is absent breeding, and perhaps a straggler or two remaining during May or June, may have led some to arrive at the conclusion it breeds here. Mr. Lubbock, a most careful observer, in his book on Norfolk birds published in 1848, mentions "a certain stream which will furnish one or two of these birds at *almost* every time of the year." Surely nearly thirty years should have proved if it did breed with us. Mr. Cordeaux quotes someone else about the young being quite a different colour from the old birds—"much lighter;" whereas the young are darker than the old birds in summer, and what he says about the gamekeeper having seen them come off from the crows' nests is not reliable. The keeper who knows the bird, probably also knows that there are persons who would give a very long price for authentic British-taken eggs. This probably would have induced him to find them if he could, and his taking them would have been a slight offence in comparison to having settled the fact of the birds breeding in England. No other account that I have seen is more definite or reliable. I have only seen this bird twice in April, often in August (young birds, with one exception), and I once killed one in February. Though I have often seen them, I never saw two rise together; they are very local, and, although two or three may be on the same stream or pool, they are generally quite apart, and they rise so soon as ever they see you, which is almost before you can see them.

Amongst the list I sent of birds not breeding in England was what I called the Common Godwit, or "Speathe" of our coast, and by which I mean the Bar-Tailed or Red Godwit of some authors, but which has not always a barred tail, nor is always red—both incorrect names. Now, excepting the lateness of this bird's northern migration, little has been said about its having

* Even supposing a bird like the Green Sandpiper has ever bred in England—the eggs or unfledged young do not appear ever to have been destroyed by man—the bird has not become more common. Is it expected to do so by protection? I say no!

been supposed to breed here, and certainly nothing at all satisfactory. Much confusion formerly existed between our two species before the curious changes of plumage were known, and I almost think some does still. Mr. Stevenson, in his "Norfolk Birds," says he has never known one **killed in June** or July, meaning the Common Godwit; but just before **he says a** correspondent of his saw some on May 9th, inland, which were tame, and he hoped they might remain to breed. Could these be **the** Black-Tailed species, which formerly did breed in England? **The present species never did, and** never will, although I have seen **the young** of it—the Common Godwit—on its first arrival, with **some of the nest-down still** adhering to it where the feathery **part of the leg joins the skin part;** this I have also seen in the Pigmy Curlew, Sanderling, and Solitary Snipe. **I think there is a misprint in Mr. Stevenson's** book respecting the date in this species, where it states "on May 26, 1862, during a severe frost."

Montagu seems to have considered **Godwits of one** species only. His description under the head Godwit refers to both; but under the head "Yarwhip" he commences by saying **it is** larger than the Godwit; he must refer to the black-tailed species, but the illustration evidently refers to the other species, a young bird, as the tail is represented **as very strongly barred** (edition 1831).

Mr. Harting, in *The Field* of April 14, mentions the two species in one paragraph. I think the **ordinary reader will scarcely understand which species he refers to when speaking of having shot the young.**

Jerdon calls the Indian bird "**Lesser Godwit.**" Now this is **wrong as it is evidently the** black-tailed species **which we see sent from India;** and, besides this, his description relates to it, **and it is considerably larger** than the **Common Godwit.** Both these **species have been** often called Red Godwits, and even their scientific names have been often mixed. Notwithstanding this, **they are most interesting birds, and** time has enabled us to distinguish **the species easily** in all their changes of plumage. It **is quite probable** the Common Godwit may be Asiatic, **but not migrate so far southwards as** India.

I think I am entitled to say something about Northumberland Garganeys, referred to by Mr Harting. In *The Field* of April. 1, 1865, under the head of "Wildfowl in the London Parks and elsewhere," (See under the head Garganey, page 5.) Garganeys certainly have always been very rare in the North of England, Mr. Harting mentions Mr. Hancock's duck, and there was one drake killed in the spring many years since at Gosforth Lake by the late Mr. W. Brandling; these are all known, but from these no conclusion that they actually bred here is decisive.

Mr. Lubbock proves their breeding in Norfolk. He mentions their appearance in March, and says those which did not intend to breed there, departed about the end of April, and he expressly states in proof of its breeding that on July 24, 1827, he shot four at sundown, three of which were young birds. This species is easily recognised when crossing you flying, the decided band on the breast being conspicuous, and the under wing coverts being half of them dark and half white, continue the band across the breast to the wings when flying. There is a remark by Mr. Lubbock which creates confusion, and I think must be a mistake, which is, that a female laid an egg in a basket in March, which seems from all we know much too early for this migratory species, as they only appear to arrive late in that month.

That this species ever bred very commonly in this country is very doubtful. Montagu, who wrote at the commencement of the present century, says it was not then ascertained whether it did. Mistakes are often made quite unintentionally. A person once told me Sheldrakes came to an inland reedy pool in spring. The bird was described as having a black head, a white neck, and chesnut breast. I asked if the duck and drake were alike, and the answer was "No!" They were Shovellers. There seem divided opinions as to the time ducks pair. I find I have several memoranda made years ago, and which have already appeared in *The Field* under the above-mentioned heading. (See under head Shovellers, page 3.) Those not intending to breed seem then to leave. Col. Montagu, who perhaps paid as much attention to birds as any person ever did, as he kept the various kinds of ducks alive to observe their changes in plumage, describes the Shoveller as by no means common.

It often happens that the birds going furthest north to breed linger longest on the road, where they can procure the requisite food, in consequence of their breeding grounds being not yet ready. The pink-footed goose* used formerly to be most common at Prestwick Carr about the middle of April. These, and many of the ducks passing northward over Holland as well as this country, at this season, will not breed till June on the desolate tracts of country which are still covered with snow, and which principally produce them in quantities. Even the Golden Plover, which breeds here in April (see Mr. Dresser's work, where nests of grey Plover are described), was found with eggs by Mr. Seebohm in Russia as late as the end of June, a difference of more than two months. These birds journeying north are then just as good food as birds killed here in February, which remain here to breed.

The different kinds of ducks, not considering the subject locally, do not all pass at the same time. During very severe weather nearly all surface-feeding ducks must leave inland places, and by far the greater portion migrate, the few which do remain being of small importance. So soon as really lasting open weather comes in spring, they begin to return on their northern journey —the Mallard and Wigeon first, the Teal and Pintail after them, the Gadwall and Shoveller perhaps next, and the Garganey last, the productions from fresh water enabling them in a great measure to live, except the Wigeon, which seem almost independent of fresh water when on the sea coast.

Many of our water birds are nearly as regular migrants as the land birds whose arrivals are so regularly chronicled; the reason why similar attention is not given to them is that they generally inhabit places further from our homes.

Mr. Fountaine's letter about the breeding of the Gadwall in Norfolk is highly interesting; but we can hardly judge from it what becomes of the flocks during very severe weather, when all fresh water must be frozen. A question arises, do they migrate?

* Breeding in the Far North.

If not, one would suppose they would stray to other places, and, from their immunity from danger at home, be easily killed. Audubon mentions that in America, this species may be readily tamed, and that it will breed in confinement, and is capable of being made useful to man. The supposition that the Gadwall was ever common in England is negatived by Col. Montagu, who states: "During the many years we have attended to the subject of ornithology, we have never been able to procure a fresh specimen of this duck.

From Mr. Lubbock's account, it would appear that the Pochard's breeding in Norfolk was not at all a common occurrence, as he specially mentions the young being met with in a single instance; and, though Col. Montagu kept these birds alive, he does not mention their breeding here. Near here I have seen them in large flocks (sometimes all males), and with tufted ducks in the middle of March, and still in flocks till the middle of April, which would appear to be the time for most of the ducks going north to breed.

I have seen the long-tailed duck remain here at sea till some of the males have acquired their singular complete summer plumage, and the Golden-eye, mature birds, apparently resting on passage till the middle of April; but no one would suggest they bred here in consequence, the former going far north, and the latter to Norway and Lapland. When we see the Golden-eye here in spring, it is generally on open sheets of water, whereas it has been ascertained to breed in hollow trees, and I should think it extremely unlikely it would breed where there are none.

It seems probable that the Golden-eyes do not breed the first year after they are hatched, as, like the Mergansers, you see the young males very far from having their fine plumage quite late in spring.

I cannot exactly see why the owner of land should not have the exclusive right to the birds, and all their belongings on it, as the owner of a lake, with a river running through it, has to the fish in it: as I apprehend so long as they are in his lake they are his, so might the birds be, so long as they are on his lands.

With respect to observations taken from live wild birds when only seen, I should like to add a word or two. There are many reasons why men write. The only reasons should be to give or receive correct information. From statements which often appear, unfortunately, there is no way to question their accuracy, unless you imply on the writer's part a desire to deceive; but when one reads a book on a subject like natural history, to study it, one must consider the probability of what is written, and not always, perhaps, jump at the same conclusions as the author. As an instance, in Mr. Gray's book on the birds of Scotland the long-legged Plover is reported as having been seen on two occasions, by Don, of Forfarshire, on mountains. Is it in the least probable one person could at different times see such a bird in such unlikely places? Another observer minutely describes the actions of one, seen with a glass at a distance, and describes it as an immature bird, even to the eyes, which were red. Does not this species retain the dark eyes like the Oyster-catcher till it is mature? If I am wrong, I willingly apologise; but it is perhaps as well to criticise statements so made, which may tend to make observers more cautious in writing.

Not long ago, in the leading daily paper, a well-known naturalist occupied a considerable space, in which he gave a quotation from the work of another naturalist on the habits of one of our common small migratory shore birds. Until I saw this, it never struck me these elegant small birds had probably received their name from their being able to *turn a stone* of fish; now, a joke is a joke, but natural history need not be brought into ridicule.*

I was delighted, on reading Mr. Gurney's book under the head "Birds claimed to be accounted British," to see he has had courage to tackle such authors as even Yarrell where he thinks they are wrong, as they sometimes undoubtedly are; but it requires a certain amount of courage to do so. I think, however, he has not even gone far enough respecting the spotted Sandpiper—I refer to the Aberdeen reputed birds. This bird seems to take the place in America which the Common Sandpiper does in Europe and Asia, most probably migrating in the same manner. Now,

* The account of the Turnstone sent by Morris from Edwards' Book to the *Times* newspaper.

I think the state of plumage of these birds should be examined; it is not stated whether they are old or young, I think the old birds of the Common Sandpiper leave us as soon as ever they have reared their young, and previous to moulting; at least I never saw one in the moult. The young leave the inland streams early in July, coming on to the coast for a few days, sometimes two or three together, and then suddenly disappear, except stragglers. I have seen one occasionally as late as October, but all these late birds are young of the year. Now in August it seems probable the old birds would have cast some of their quills, and be quite incapable of taking any such flight as across the Atlantic; and the Spotted Sandpiper does not, I believe, go to the far north, as several of the American Sandpipers which occasionally visit us do; and it is more likely these birds reach us along the shore, taking from some accidental cause the European for the American continent, and, having once made the mistake, come straight on. The circumstances of these two species correspond with the Woodcocks of the two named tracts of country. Another most unlikely part of the whole business is that two should have been together, as this tribe of Sandpipers are generally solitary in their habits except at breeding time. If it is suggested they started together to cross the Atlantic, it is even a marvel of bird travelling that they should have been able to keep company for such a distance, and I must say I cannot believe it.

Two of the very limited number of the British Buff-breasted Sandpipers are recorded by Yarrel and others as having been obtained at different times by one man. This appears remarkable. After long experience, one can nearly judge from a British Sandpiper's appearance to a few weeks of the season it was procured. It would be satisfactory if some experienced naturalist would go further into this, and examine the two birds, and give his opinion whether their plumage corresponds with the different seasons they were said to have been killed. This might be done perhaps by comparing them with skins from America in the different states of plumage—that is, the young bird after being fledged, the winter plumage, and that of summer—and to trace if the changes corresponded with the dates of reputed capture.

Is not more notice taken of the sale of Plovers than necessary? At the price they are sold in London, it would never pay to get them unless they were taken in quantities, as I believe they are in Holland by nets, and then only when they are in flocks. The extent of country inhabited by the Golden Plover and Peewit is immense. The former probably breeds over the hilly parts of Norway, Sweden, Lapland, Russia, and Iceland, the greater portion of which is desolate waste; and when the millions there are of them are taken into consideration, what does the quantity captured amount to? Merely nothing. If they are of so much importance, why not bring in an Act of Parliament to prevent the reclamation and improvement of the bogs and waste lands in this country, on which they ordinarily breed? The extent of country over which species are scattered varies much; but a bird like the Teal, which appears to be common in Europe, Asia, and in winter in Africa, and even in America, is represented by a closely-allied species, if not only a permanent variety*—of what consequence is it that many of them are killed? By all means give every one the power to protect these on his own land and water; but we certainly should not prevent our neighbours the foreigners from sending their produce to the best market under the circumstances.

A drake Teal in March, when clean and unruffled, is simply a gem, and how few know it! nothing can surpass its beauty. When I can easily procure one, I often have it for a few days to look at, and I have sometimes shown one to others, who have been surprised at it, and would hardly believe it was an English bird. And what can exceed the chasteness of the plumage of a fine drake Garganey at the same season—the exquisite arrangement of lavender and white colours, and the green and brown? If we have a complete close time from February 15, no one can ever again have an opportunity to see one in England, as they never come even to Holland till late in March; and even if they could be procured in autumn, which I much doubt, not a single bird would be in full plumage.

Living in the northern half of the world, we are always

* I observe the illustration of the Teal in one of my copies of Montagu's work is taken from the American bird.

accustomed to observe all birds migrating north in spring to breed. Now, I wonder if any species in the southren half migrate southwards to breed in what would be our autumn, but to them spring, as in August, September, and October, which would be towards the Antarctic Circle. (On looking at a map of the world this probably appears unlikely, as there appears to be so little land free from ice southwards where they could breed, in comparison to the land about the Artic Circle.) If there is a southern migration of species inhabiting the southern portion to breed, they would be acquiring their summer plumage in our autumn, the very reverse time to that when our northern birds do; and in cases where the two sets of migratory birds meet (if they do), the contrary changes in plumage would completely puzzle anyone. Perhaps some species of shore birds inhabiting only the southern half of the world do not get a summer plumage.

I suppose that such birds as the Pigmy Curlew and Grey Plover—the latter migrating to Australia, and the former throughout the Indian Archipelago (Jerdon's "Birds of India")—return within the Arctic Circle in Asia to breed; but for what cause do they extend their migratory flight so far southward? It cannot be for food only.

I think I have heard or read that these and similar species, when met with so far south, are never found in their summer plumage. It would, therefore, be useful to know the precise dates when they have been got so far south, to ascertain whether they all do not again come northward so soon as the approaching breeding season is about to commence.

There is another point to which attention should be given. In these day of rapid transit, it is very difficult to say what distance recently killed birds may be brought; and collectors should be warned not to give excessive prices for birds said to be killed in England or Scotland, unless they get them from very reliable sources. Nothing is easier than for a Snowy Owl or Eagle to be sent from Sweden or Norway along with grouse to some eastern port in Scotland or England, and to be procured there and forwarded to some other person at a distance with a false account of its capture, which may be apparently reliable and minute.

Even now, April 20, a quantity of Grouse, Capercailzie, and Black Game have just arrived here in very good condition. How long they may have been killed I do not know, as I believe they are collected together and kept frozen; but the White Grouse have not commenced to get their breeding plumage. The Black Game, as usual, are to an accustomed eye a different race to our own—the cocks, as a rule, larger and darker coloured, and with larger tails; the hens invariably much darker than our British examples, and the markings on the back in many much more resembling the hen Capercailzie. Norway is a totally different country to this, and the ground over which the game is bred is so unbounded in extent that in the breeding season, and after it, the broods must be far more scattered and difficult to meet with. It is, I believe, in the winter the birds are driven towards the villages by the snow, and at that time become useful to the inhabitants. It will be a very long time yet, if it ever can be, before the country is populated sufficiently to reduce the numbers of the birds, even suppose all the gunners from Scotland and England were put down in the country for a season or two. So soon as the snow goes, the birds go back to the mountains again, and then it would not pay to follow them, even to send them to the English market.

Migratory birds being met with, either here or on the Continent, till the middle of April, and even very much later, should not lead us to say they breed where they are seen, unless we have positive facts to guide us. It is only at this time our ordinary summer visitants, as the Swallows and Warblers, arrive here. Now most of the same species go very much further north on the Continent, and, as with the other breeding birds, if they went too far north too early they would be starved. Many of the birds we see here till even June are only taking their part in the extensive vernal migration northward, which prevails all round the world, and in which a far greater number of our species take part than is generally supposed; and the further north we go, the later the arrival of the several species will be which intend to make the north their summer home.

CHARLES MURRAY ADAMSON.

MIGRATIONS OF BIRDS.

(Reprinted from " The Field" Newspaper, Sept., 1866.)

Can any of your readers tell us about the **migrations of the** Knot, Common Godwit, Grey Plover, Sanderling (and even the Dunlin, though it differs from the others in breeding commonly in some **parts** of our Island)? All these species arrive early in autumn **on** our coasts, indeed, as early as the young can fly well; and although, except the Dunlin, none of the others breed beyond **the** limits of the arctic circle, I have had the young of them **all, with a** portion of the down that they got on being hatched **still adhering to parts of** the body. From the time the young **birds arrive, these** species are met with in greater or less numbers, sometimes in vast quantities, during the whole winter, braving the severest weather; but no sooner has the earliest spring appeared, than by far the greater portion of them leave our shores entirely before commencing to get their summer plumage. Now, we should expect that, as spring comes, they would repair towards their breeding grounds. But do they? What I wish to know is where they spend the time that elapses between March and their appearance in May, when they arrive **in greater** or less numbers on various parts of our coast every year, **many of** them in their full summer plumage. I think their appearance in May is when they are actually on their way to their breeding grounds, as, if they arrive at them by the middle of June, they would be early enough for the situation, and still have sufficient time to rear young, which would be in the state we find them in early in September. We should expect, if they did not appear again later in the season, that they had gone further north, and were nearing their breeding grounds as the weather became milder; but their later appearance in their breeding plumage renders this supposition unlikely. But where do they all spend the time they are undergoing the change of plumage and are absent from us—that is, if they are the same flocks which wintered with us? On the 4th of August I had a Green Sandpiper sent to me. It is ten days earlier than the young birds usually arrive: it was, however, an old bird, having nearly lost the freckled with white plumage on the upper parts, and the quills and tail were moulting. It is the first time I have noticed an old bird returned immediately after breeding; and on the 14th of August I saw two Greenshanks, evidently passing on their southern journey, as they were not at the same place on the following day. On imitating their wild cry one came very near, flying with the usual vigorous flight of the species. Their flight appears nearly as superior to that of the Redshank as that of the Swift does to the Martins and Swallows. It is easily known on the wing from the Redshank by the absence of white on the quills: indeed, the wings of the Greenshank look quite black, and contrast well with its white body, particularly as it is seen when alighting. Near here, at this season, the Greenshank is as often seen near shallow pools of fresh water and at a distance from the sea side as near it: later on in the season, when it is seen, it is always by the sea side. On the 6th of July I met with the Redshank breeding on the Northumberland moors, thirty miles from the sea: the young were capable of flying short distances only, the quill not being of their full length.— C. M. A. (Newcastle-upon-Tyne).

MIGRATORY BIRDS.

(Reprinted from " The Field" Newspaper, June 7th, 1862.)

You have many notices of the appearance of the usual migratory small birds. It is curious to observe how nearly some of them come to a particular day each year; equally curious is the mysterious manner in which they do come. Perhaps the first appearance of a species is observed where you would not expect to hear of it; then several notices of its appearance are recorded in various places; but how do the enormous quantities of Willow Wrens, Swallows, and other warblers, get dispersed over the country as they do in their unobtrusive manner? You go out on an April morning and find every plantation full of Willow Wrens, where there was scarcely one the day before. Has any one observed them travelling? Do they come by night, in large flocks, or singly? The numbers of one of the common species alone dispersed over the country must be surprising, and as there are so many species on the move about the same time, one would expect to see or hear of large flocks or single birds being seen continually arriving; but one does not, they appear scattered over the country all at once, at the regular places suitable to them. Fortunately, this great army of small insect-devourers is not likely to be diminished in numbers by poisoned grain.—C. M. A. (Northumberland.)

MIGRATORY BIRDS.

June 6th, 1863.

A SHORT time ago an inquiry was made whether the Starling was migratory. At first sight this appears to be a question easily answered, but it is not so. Are not a far greater number of the species of birds we have migratory than is generally believed? Before determining whether a species is migratory, we must not only look at home, but endeavour to find out its habits abroad. We will then probably find that although it may be found throughout the year in this country, it only appears in others at certain seasons. The Skylark and Common Thrush are not generally supposed to be migratory, but when a severe snowstorm comes early in the season and remains for a length of time, do they not migrate from this country? Indeed, is it not possible that we get a great many, if not nearly all we have in winter, from a distance every winter? We are told that these species are migratory on some parts of the continent, and therefore if they are enabled to foretell winter there, why need we doubt their being able to foretell a severe winter here, and be able to leave the country before they are overtaken by a storm which might probably destroy them? It is interesting to hear accounts of the different sort of birds which alight on ships or are seen at a distance from land; but notices of their appearance should be received with caution, as they may come from persons who do not know the various species sufficiently. Supposing the Starling has two broods in the year, unless a great portion of them do migrate and do not return to this country, they would increase in numbers far more than they do. In autumn, when in large flocks, and they fly about in their singular manner, may they not be preparing for a long flight? They often fly and wheel about for a considerable time in the evening before settling to rest, and at that season there certainly appear to be in places many more than when they return to their breeding places in the early spring, even taking into consideration that they are then more scattered about. In the autumn most water birds get very fat; this seems to be a provision of nature to enable them to withstand the cold and scarcity of food in case of severe weather. Nature has probably given birds the power of knowing when a severe storm is approaching. How few birds, both in numbers and species, particularly land birds, are to be seen after a long continuance of hard weather early in winter. In open winters like the past some birds, as the Peewit, are never totally absent; but when a severe and general storm comes early and lasts a long time, do such birds not leave the country entirely, but return immediately on the weather becoming mild? Many species probably migrate backwards and forwards as they can procure food, as the Grey Geese, Teal, Plovers, and others. As to their being able to fly long distances in a very short space of time there cannot be any doubt. I think hard weather presses birds most when it has not been very severe during early winter, so as to drive them away; but about February, when a great portion of their autumnal condition is lost, if they are then caught by a severe storm they seem not to be inclined to leave the country, or probably wait on, expecting mild weather till they are too weak to get away, and sometimes get very much reduced, and perhaps perish, but Nature has taken care generally to have sufficient of each species in different localities to make up for all casualties. For instance, if it were possible for all the birds of the most of the species we have in some particular tract of country to be killed either by storm or failure of food, others of the same species would probably speedily make their appearance if the country be adapted for them, and presently occupy the same ground. There seems to be no rule for the migration of birds. Some species, the Arctic Tern for instance, is a summer resident only on our coast: it appears to be the same in Greenland. If it can procure food and what is requisite for its young, one wonders why it should travel so great a distance apparently unnecessarily. In some places the Common, Sandwich, and Roseate Terns associate with the Arctic Tern while breeding, and seem to require the same food and situation for their nests. Why do they not, also, go so far north? How is it that the Little Auk, when it comes from its northern haunts in autumn and early in winter to these shores, is generally found dead or dying—at any rate, very lean and exhausted? where is its regular winter home, where it must be in millions? What becomes of the Phalaropes in winter? I have never seen the old Red-necked Phalarope in its winter plumage, a bird of the year only appearing occasionally in autumn: the Grey Phalarope, also, only occasionally pays us a visit during his autumnal migration, and not one in fifty is an old bird. Where can the hosts of Knots, Common Godwits, Grey Plovers, and Sanderlings, which sometimes visit us, breed in Europe? The breeding grounds of the Knot, Grey Plover, and Sanderling are known in North America, but the Common Godwit is not an American species, though it appears exactly at the same seasons and manner as those other species last mentioned, on our coasts. How do the young birds of the year only, of many species, find their way to us in autumn, as the Great Snipe, Pigmy Curlew, Little Stint, Temminck's Sandpiper, Green Sandpiper, Spotted Redshank, and others? Why does the Black-headed Gull sometimes depart from its usual custom and breed by the sea side, and the Lesser Black-backed Gull sometimes leave the sea and breed on the moors far inland? The distribution of species is curious. In many instances, one is found in particular districts and at particular seasons only, coming with great regularity, and rarely observed except at those places, whilst another nearly allied, gets generally distributed throughout the country. One often hears a great outcry about a bird having been killed, as if the species really suffered to any extent by its death. Those who make the outcry apparently do not remember that the most beautiful and interesting species—be it bird, insect, or flower—was not intended to, nor does it, escape contributing its share for the benefit of others. If the bird be a migratory one, it had to run the risk of being lost either in a gale or fog, or, perhaps, when fatigued after reaching land, falling a prey to a Hawk or some other natural enemy. No doubt, destroying a pair of birds which have come to breed at a particular place, you and others may, perhaps, be deprived of the pleasure of hearing and seeing the species for the season; but the individuals killed are of small account in estimating the number of the species. Few species, I think, are really rare, that is, few in number. I apprehend the difficulty is to find their peculiar locality, and at it they will not be uncommon. The reason of this supposition is that, judging from a bird like the Grey Plover, it is probable, from its being a numerous species, its breeding ground will extend in Europe over a very great extent of country, like the Golden Plover; and if this is so, and its breeding ground is yet unknown to us, there must be plenty of ground yet unsearched where many species that appear rare to us may be abundant. The exceptions to this seems to be such birds only as have not the usual means of escape, namely, wings. Where such birds existed, and were confined, perhaps, to a limited district, and there killed out or nearly so, there can be no source left from which to replenish the depopulated country. Many complaints are made of the few wild ducks in comparison to what there were formerly. The reason in a great measure probably is that nearly all the small strands, backwaters, and shallows, by the river and lake sides, where they used to feed at nights, are drained, and consequently do not hold water long enough to breed the insect and shell, and grow the plant food on which they feed; and the birds, being deprived of their feeding grounds, do not remain in the neighbourhood, even though the main water is left untouched. I have derived the greatest pleasure from seeing birds alive and wild, and also from sometimes getting hold of a recently killed bird, the shape and colour of the legs, beak, and other parts, in some species, look so fresh and beautiful; and without having dead birds, it is not possible to learn anything about the changes of plumage, food, and other matters. No one, however, regrets more than I do their wanton destruction, and their being driven away by what are so often termed improvements in the country. When we are informed that some of the smaller species of humming birds scatter themselves over a great portion of North America during the summer and retire before winter after breeding, why need we be surprised at the powers of flight the larger birds have? C. M. A.

BRITISH SPECIES OF BIRDS IN INDIA AND AMERICA.

(Reprinted from " The Field" Newspaper, Sept. 22nd, 1865.)

SIR,—On looking over Jerdon's "Bird's of India," I find so many of our common species that I have been tempted to make a list, for comparison, of those found there; and, having done so, I have marked with a † those met with also in North America. Perhaps this list may interest some of your readers. The number of species, the same in India as with us, appears to be about one hundred and ten, and of these nearly forty are also met with in North America. I have inserted a few species where either a permanent variety or a most nearly allied species represents the one in another country. Those in the list are chiefly migratory birds and chiefly water birds, and their appearance in India corresponds generally with that here, as well as in America. Now I must observe that I do not wish anyone who may read this to suppose for one instant that it is anything like correct; there is no one living who could make a strictly correct account, nor will there ever be. The subject will occupy man's attention for ever, as there must always be diversity of opinion as to what constitutes a species and what a permanent variety.

LIST OF BIRDS FOUND IN INDIA WHICH ARE BRITISH SPECIES
(those marked with a † being also found in North America).

†Peregrine Falcon.
Hobby.
Merlin.
Kestrel.
†Goshawk.
Sparrowhawk.
†Golden Eagle.
†Osprey.
†Common Buzzard.
Hen Harrier.
Ash-coloured Harrier.
Moor Buzzard.
†Long-eared Owl.
†Short-eared Owl.
†Common Swallow.
†Sand Martin.
Swift.
Bee-eater.
Roller.
Wryneck.
Cuckoo.
Hoopoe.
Fieldfare.
Redwing.
Wheatear.
Redstart.
Willow Wren.
Lesser Whitethroat.
Treelark.
†Raven.
Carrion Crow.
Rook.
Jackdaw.
Starling.
Rose-coloured Starling.
Tree Sparrow.
Ortolan.
Black-headed Bunting.
Mountain Finch.
Quail.

†Grey Plover.
Kentish Plover.
Peewit.
Norfolk Plover.
†Turnstone.
Oyster Catcher.
Solitary Snipe (distinct or larger variety than European).
Crane.
Woodcock (represented in America by another totally distinct species).
Snipe (represented by very nearly allied species, if distinct).
Jack Snipe.
Curlew (represented in America by distinct species).
Whimbrel (ditto).
Black-tailed Godwit (called by Jerdon Lesser Godwit).
Ruff.
†Knot.
†Pigmy Curlew.
†Dunlin.
Little Stint (nearly allied species in America).
Temminck's Stint.
†Sanderling.
†Grey Phalarope.
†Red-necked Phalarope.
Wood Sandpiper (Glareola; called by Jerdon Spotted Sandpiper).
Green Sandpiper.
Common Sandpiper (represented in America by Spotted Sandpiper, closely allied, but quite distinct).
†Greenshank.
Redshank.
Spotted Redshank.

Longlegged Plover.
Avocet (represented in America by closely allied species).
Coot.
†Water Hen.
Spotted Crake.
Black Stork.
Stork.
Heron.
Purple Heron.
Little Bittern.
Bittern.
†Night Heron.
Spoonbill (represented in America by totally distinct but closely allied species).
†Glossy Ibis.
Common Wild Goose (origin of Domestic Geese).
Pink-footed Goose.
†White-footed Goose.
Ruddy Sheldrake.
Sheldrake.
†Shoveller.
†Mallard.
†Gadwell.

†Pintail.
†Wigeon (? permanent variety in America, or distinct species).
†Teal (? permanent variety in America, or distinct species).
Garganey (represented in America by the Blue-winged Teal.
Red-crested Pochard.
†Pochard.
White-eyed Pochard.
†Scaup Duck.
Tufted Duck (represented by a nearly allied species, with brown band round neck).
†Goosander.
†Smew.
†Crested Grebe.
Little Grebe.
†Great Black-backed Gull.
Lesser Black-backed Gull.
Black-headed Gull.
†Gull-billed Tern.
†Common Tern.
†Little Tern.
†Cormorant.

Many of our other species seem to be represented in India by closely allied species, as the Cushat, Rock Dove, Stock Dove, Golden Plover, Solitary Snipe, and others. I have sometimes had opportunities of seeing skins of some of these species from India. The wading birds have been generally in their mature winter plumage, and from that changing and changed to summer—no young birds in their first plumage being amongst them that I remember.

It seems difficult to account for the absence from India of some of the eastern European species, which either do not, or rarely appear westward. We might expect such species to be more likely to occur there than some of those which do, and which are equally common on the American continent, appearing there according to the seasons, as in India and with us. Such birds as the real Bimaculated Duck and the Red-breasted Goose may very likely spread over a great portion of Northern Asia, and yet not migrate so far south as India. There can, however, be little doubt but that some species are distributed laterally to a much greater extent than others. If anyone takes a map of the world and looks at the extent of country inhabited by a species, which occurs in all the four quarters of the globe, and probably breeds in three of them—Europe, Asia, and America—he may consider the numbers it is likely to consist of, and the little fear there consequently is of any appreciable harm being done in killing, you may say, any quantity you can.

The summer residence of a bird is properly its home; thither it goes to perform the office nature intended it. Probably throughout the northern portion of the world a similar migration is taking place at the same season, the unbounded and thinly populated swamps, and wastes, and forests during summer affording food and shelter for myriads of old birds repairing to them at that season, and which (probably multiplied on an average threefold by their young), on the advance of autumn again scatter themselves to obtain food, as well as to form food for man and other animals. The same process has gone on since the commencement of time, and will continue to do so as long as the world lasts. Though some birds are summer migrants with us, and others winter migrants, the same course is pursued, and for similar purposes; the only difference is, that our summer migrants are more tender than those of winter. It would appear that those species whose visits are least frequent are not intended to be our companions, and consequently their destruction is of little importance. If the species is intended to be useful in a country, it establishes itself in spite of all opposition. The elasticity, if I may so call it, of species is surprising. Who could estimate the number of Skylarks used annually in London alone? And yet the species is as plentiful as ever, and that without man's aid. Nature's supplies appear to be without limit in most cases.
C. M. A.

STRAY THOUGHTS ABOUT BIRDS—THEIR MIGRATIONS, ETC.

(Reprinted from "The Field," October, 1865.

THE Birds of England might be divided into four imaginary groups—first, those which remain throughout the year; secondly, those which come to us for the summer; thirdly, those which come for the winter; and fourthly, those which visit us, but remain **neither summer nor winter.** And the fourth group **might be again** separated into **those** whose visits are periodical, **and possess some degree of regularity in their passage** to or from **their winter** and summer haunts, and those which **come only at** uncertain and irregular periods.

Though I have made four divisions, it would be impossible to class certain species in any particular one of them, as many **birds** would have to be considered, perhaps, as belonging to more than one, in consequence of the different counties where observations might be taken; besides which, it is not possible to know whether the same species found at different seasons is composed of the same individuals. We, however, certainly **have a** great number of species. Generally speaking, the further north such observations are made the more summer visitants are found in comparison **to the** number of species; and when you get to the **extreme north all will be summer** visitants, with few exceptions **indeed.**

With respect to residents, it seems by no means **certain that** many species which are found at all seasons in some counties, and which are generally considered to be residents, are entirely so. **Some** certainly appear to **be, as the** Red Grouse and Black Grouse; but is the Partridge strictly so? Is it **not** migratory **in some countries**? It appears in the north of England in severe seasons either to leave the high pasture lands, or be killed off, and it does not appear to get dispersed over the ground again unless several mild seasons occur in succession. Moreover, does it not sometimes cross the Channel?

With respect to the second group, so much has been written about the periods of arrival and departure of the commoner kinds, that I will pass them over, only remarking how little is known about their proceedings during their absence from us, and the distance travelled by them between the time of their leaving us and their return.

The third group comprises many species, some of whose summer haunts are pretty well ascertained; but it also comprises other common species, whose summer haunts are still but little known to us, as the Knot, Sanderling, Common Godwit, Grey Plover, Smew, and many others.

The first portion of the fourth group comprises several species which formerly were of the second group, but whose haunts and habits with us have become changed by man's work; such are the Bustard, Black-tailed Godwit, Avocet, Spoonbill, probably the Black Tern, and others. The breeding-grounds of each of these species appear to have been very limited in extent in this country; and those places having been destroyed, and no others suitable for the birds being at hand, they have in consequence retired elsewhere; but probably as species they are as numerous as ever they were. This group comprises many species which appear to have the widest range of flight; some of the species which probably breed in the far north merely pass us and spend the winter elsewhere, as the Pigmy Curlew, Little Stint, the Phalaropes, Buffon's Skua, Pomerine Skua, Little Gull, and others; whilst other species breeding a much shorter distance northward, although strictly migratory, belong to the third group, as the Turnstone, Purple Sandpiper, and several others.

I have often wondered that the Black-tailed Godwit is not more frequently met with in the north of England and Scotland. It appears to be a common summer bird in Iceland, but is not known in America; consequently we might expect to meet with it on its passage to Iceland to and from its winter quarters, which are apparently in much warmer climes than our Island. Now I never knew an authentic instance of its being seen on our north coast in spring. I have known it killed in July in a few instances, old birds apparently having bred. I have also on one or two occasions met with young birds killed in August on the Solway shores. Whether these are stragglers from Iceland it is not for me to say; **but** what becomes of the regular flights from that country? It is quite different with the Common **Godwit**—it is not known to breed in Iceland, neither is it known in **America; but in** September the young birds are sometimes com**mon, and the** species is to be found all through the winter and till the spring months, some getting **much** spotted with red before leaving in April. **Occasionally we even** see them **in** red plumage, apparently stragglers **from the flocks** which appear on the south coast in May, on their **journey to** their unknown breeding quarters. Old birds **when red are** rarely seen in **early autumn.** Now, this species seems to migrate **in autumn from east to west, whilst the other** appears to migrate from **west to east;** but so little appears to me to be known at present of the migrations of birds, that the more one thinks about them, the less he finds he knows. The Knot appears to migrate in the **same course** as the Common Godwit; still the former is equally common in America and here, whilst the latter is unknown there. I have shot a Knot returned from its breeding ground as early as July 19th, in its full summer plumage. In the early part of August a few old birds generally occur, much faded in plumage. It **is** curious **to observe** how greatly the plumage fades between the time of their departure for breeding **and** that of their return; by **the** latter end of August the young arrive. The species is to be **found almost** invariably by the sea-side throughout the winter, **and sometimes as late as** May; but I never saw one killed here **(Northumberland)** in the red plumage they get before departing **to breed, and in which** they are met with on the south-east **coasts.** I have seen them in May here without any red feathers **at all.**

Migratory birds remind me of what is said of the wind—we hear the sound, but cannot tell whence it cometh or whither it **goeth;** so it is with them—we see them, but are unable to tell whence coming or whither going. Some of the species of birds enumerated as English may have escaped from captivity, and being afterwards shot, have been described as British. Some may have been described as British in other ways—the Spotted Sandpiper for instance, which seems in the new world to take the place of the Common Sandpiper in the old. Was the Spotted Sandpiper ever thought of as British till Bewick's figure of a Common Sandpiper, with the wrong name, appeared? and the bird, having thus been admitted as British, has not some over-anxious bird-stuffer made American specimens British for profit's sake? When it is considersd how rarely correct information is obtained in such matters, such a suggestion as I have made may not be far from right. I have no object but a desire to have facts properly represented, neither to make species commoner nor rarer than they really are. When I collected birds I would rarely keep one unless I had the bird before it was skinned.

Questions are sometimes asked, in consequence of the early appearance of a Woodcock, if it was likely to have bred in the neighbourhood. This year during the fall (October 10th) several Woodcocks have been seen and shot on the coast—you may say, during their migration, as I believe they are sometimes watched for and shot before actually alighting. But an opinion as to where they were bred must, it seems to me, be the merest conjecture, as their breeding-grounds, when suitable localities occur, probably reach from Western Europe to Eastern Asia. I do not think that any Woodcocks bred in England are at all likely to remain over winter, or even Snipes. Woodcocks are sometimes met with in August; these, probably, are stragglers, which have bred, or been bred, in England or Scotland, and are then on their journey southward. It is impossible to form an opinion of the migrations of a particular species from one spot, or to form a general idea from one species; but though each species has its peculiarities, there is generally some similarity in the habits of nearly-allied species. The Common Snipe appears in such places as suit it during the breeding season, and for a short time only after the young can fly well they are to be found in the vicinity —the old birds remaining until the wings are again perfect, which is about the beginning of September. After this time the neighbourhood seems deserted; but as winter advances these birds are again dispersed over the country in the same manner as the Jack Snipes, and I am inclined to think that, like them, those then found come from a distance.

I have seen in *The Field* recently the remarks about the eggs of the Jack Snipe. My impression on reading them was that it was a Quail's nest that was found. Has the Jack Snipe ever really been detected breeding in England? All nests said to have been of it appear to have been those of the Dunlin. Why the Jack Snipe does not breed in England, while the common one does so commonly, is one of the mysteries we cannot fathom. So far as we know, it seems to be a less numerous species, and to have a more limited range than the Common Snipe. It would be quite impossible for the Jack Snipe to cover nine eggs of the size it lays; indeed, it is only by the singular way they are placed in the nest that it can cover its four.

With respect to the Woodcocks of the old and new worlds, you have the two species totally distinct, and their districts distinct; each species is as typical as the other, and each as far removed from any other genus or species. I saw in the Exhibition of 1861 a photograph of a European Woodcock which had been found dead (if I remember right) on the Labrador coast; and it was stated to be the only instance of its occurrence in America. That some species, as the three Divers, are the same on both continents, and others, as the Woodcocks, though totally distinct, yet nearly allied, and apparently taking the place of each other, is very singular. The only three known species of Divers appear to be equally common in the northern regions, on the continents of both the old and new worlds, and to have similar habits; each of the species is as typical of the genus as the others, and none of them appears to be more nearly allied to any other genus than the others; and though the three species are closely allied to each other, how wide is the distance, comparatively speaking, between them and any other genus. In fact, it is not easy to connect any other bird with them. In many respects how much they differ from the Guillemots and even the Grebes.

It appears to me the original framers of the game laws understood the uncertain appearance of the migratory birds (which they omitted as game, but which required a game certificate to follow) better than the generality of people do at the present day—such are Quail, Snipe, Woodcock, and Landrail, which were wisely considered, like wildfowl, as belonging to anyone on whose ground they were for the time. The Quail is generally considered a summer bird, and probably migrates regularly in spring and autumn. The only instances I have met with it alive and procured it were twice—once on the 1st of January, and once on the 3rd of the same month, in different years—both birds being very fat, and their crops full of the seeds of a weed commonly called fat-hen.

It is difficult to account for some species not increasing rapidly in number. Instance the Landrail and Starling. The former comes abundantly the first week in May in each year, and lays from eight to ten eggs. Many undoubtedly are destroyed, but still the destruction of the species here can be nothing in comparison to those reared in safety, and which quietly and imperceptibly leave the country previous to the crops being removed from the ground; but the species does not apparently increase as one would expect it to do under the circumstances. The Starling comes to its regular breeding places in spring, and lays about six eggs. The number destroyed certainly is nothing to be compared to the increase there is in autumn, at which season you meet with flocks of hundreds, but which disappear before winter, and the next spring only about the ordinary number come as before. I fancy there must be some active agent destroying them in other countries that we know nothing of, as decidedly far more go away than ever return. Similar remarks might be made about many of our summer visitors. Apparently the numbers that come the following year have little reference to the previous year. It would appear that so many are bred, so many are required, and still the number requisite to replenish the stock is always forthcoming. It is not as if England alone monopolised a species in summer: the migration of each species probably, and more frequently than not, extends from east to west of a quarter of the globe. It even seems doubtful whether the destruction of some of the larger species of migratory Hawks in England really lessens their numbers. Many of them are only migrants here, merely as stragglers, periodically passing from the vast tracts of **ground where they** spend the summer **unmolested**, and probably each species never was commoner in England than at the present time—for instance, the Goshawk, **Rough-legged Buzzard**, Honey Buzzard, the Jer-falcon, the Osprey, and others; and I even think that if these birds were not interfered with when **they** do visit us, no perceptible amount of injury **would** ensue, as they are all wanderers, **and seldom tarry** long at one place.

C. M. A.

THE WOODCOCK.

Reprinted from "The Field" Newspaper, Dec. 27th, 1862.

THE number of Woodcocks in many places this season will perhaps satisfy some persons that there is no probability **or** possibility **of** the species being exterminated. It seems impossible to account for their migrations, but it appears to be now ascertained that, like the whole of the tribe of Sandpipers and Plovers, they leave the places where they have been bred soon after they are well able **to fly**. Why the majority of those which come in autumn leave **certain woods** soon after their arrival, **when, perhaps**, during the whole winter **a few may** be found, **and those** probably very fat, is difficult to **say**: it seems to show that it is not want of food that takes them away. The only way to make out whether there are different **races of** Woodcocks would be to visit their **breeding** places during summer, and ascertain what the **pairs of birds are** like which are breeding together. It may happen that the flights are sometimes composed of birds of one sex only, and that would account for large birds, probably all females, being sometimes found at one time, and smaller birds, probably all males, being found at another time. I have given very great attention to the changes **in** the plumage of waders, and consider that after the general autumnal moult, the year after the birds are hatched, it is not possible to distinguish the age of a bird of this tribe by its plumage. At or previous to that time the bird loses all trace of its first feathers, and the change afterwards is regular from winter to summer; and surprising in some species are those changes. No two things in nature are exactly alike; doubtless every intermediate state between the extremes, both in weight and colour, are to be found. I have an accidental variety; more than three-quarters of the bird are as white as a Ptarmigan in winter.—C. M. A. (Newcastle-on-Tyne.)

ON THE APPEARANCE OF THE SAND GROUSE IN ENGLAND.

(Reprinted from " The Field" Newspaper, June 20th, 1863.

EVERY one, whether fond of Natural History or merely of shooting, who has read the notices of these birds having been shot and seen at the various localities in England, must be astonished at their having visited us. When we consider the part of the world ordinarily inhabited by them, their appearance here is really surprising, and it makes one desirous if possible to try to account for it; had it been the European Sand Grouse, and its progress towards us had been gradual, coming nearer to us year by year, we might have expected that it would ultimately have acclimatised itself, or at any rate have paid us occasional visits; but the immense space between the countries inhabited by the present bird and our island, seems to render its either remaining with us or visiting us again, perhaps for very many years, uncertain and even improbable. The only supposition we can arrive at is, that some flight or flights (probably during a regular migration) have been by some accidental cause utterly lost, and have by mere chance reached our Island. Though it is called a Grouse, it is very dissimilar to any bird of that tribe we have; the formation of the wing is unlike that of any of our species, the secondaries are short and the primaries rapidly increase in length, showing that it will have a different flight from them.

It is probably a bird intended by nature to take very long flights over huge tracts of desolate sandy and barren country to procure food, and it seems scarcely likely that the present enclosed state of this country could suit it; however, from the quantity seen, and the extent of the country where they have been, there must have been many flocks which have not been met with or disturbed at all, and it will be fair to see whether any do **remain**. Their disappearance **will** probably be as unexpected as their appearance; it seems very doubtful whether any of these (I call them lost) birds will ever reach their native country again. If no birds **were** to be shot we would probably have remained in ignorance **of** the interesting appearance **amongst us** of these strangers. Fortunately, birds are not easily driven from ground they have a liking to. It is a difficult matter to drive Partridges from a particular field; they become wilder by being disturbed and repeatedly shot at, but they are not easily made to forsake the field entirely, except perhaps by removing the crop on **which they** either feed or amongst which they find shelter.—C. M. A.

(*Reprinted from* "*The Field*" *Newspaper for August 8th*, 1863.

SIR,—I hope **the severe remarks relating** to the killing of Sand Grouse will **not deter your correspondents** from giving further accounts about **them; and I hope that** any communications **you may** get about **their moulting, nesting,** or departure will be such as may be relied **on. We have as** yet heard little which leads **us to** believe **that** they came **here** for the purpose of breeding. **No** one has told us that they had paired, and the pair had separated from the flocks; should they not be discovered breeding **here,** the fact of the number having been shot cannot reasonably be **given as the** cause **of their** not having done so, as there must have been numbers **of flocks** unobserved and undisturbed; but supposing **every** flock **had been** seen, and that one half of the number in **each flock had been** killed, that would not have prevented the **remainder from breeding** if they had come here for such purpose. Birds **cannot tell** each other that they will all be killed if they remain, and advise their companions to go where they will be better received; they were probably as likely to have remained at Heligoland as with us. Professor Gatke, in his letter, does not wish any of your readers to suppose that the birds which he enumerates as having been met with, if they had not been killed, would either have remained to breed, or have returned the next season; they appear there by accident, the individuals having missed their regular migratory course. I for one am glad that they have been met with, and that notice of their having been found has been given as he has given it. It is oftentimes very difficult indeed to make out the sex of birds, even after careful examination; and any information on that subject got from an ordinary bird-stuffer should be received with caution. We hear much more about the destruction of birds at the present time than there is occasion for. No one regrets more than I do their being killed for no purpose; but to try to discourage young men from the pleasure of blending the sport of shooting and the study of Natural History together, is to deprive them of a most useful and amusing occupation. Much information would be gained if more of them would follow such a pleasant pursuit as Ornithology, and in which so much has yet to be learnt; and the time they would devote to it might be very much worse spent. Not only so, but I maintain that all things in nature were given to us to study and admire, and that there is no harm in destroying animals for the purpose of studying nature in all its branches, notwithstanding opposite views are taken by so many persons. Surely the most miserably-stuffed bird in a poor man's house, if he chooses to have it, is as useful as a bird shot by a rich man for the purpose of being eaten; or one of those which are commonly called vermin, killed, and nailed against a wall.

I cannot believe that any amount of protection will induce birds to breed regularly in this Island which only occasionally visit us, and which are not accustomed to breed here. I saw this spring a notice of a Hoopoe having been shot, which on examination showed that it was a breeding bird. At the time I regretted that your correspondent did not show beyond a doubt that **it had a nest**. **If he** had gone again **to** the place he would **most** likely have seen the other bird, or **at** any rate he might **have** heard of it; he might have ascertained whether there was a likely place for the nest, and have sought day and night **till** he found it, and cleared up the matter satisfactorily. I would be **the** last person to disturb **a Hoopoe at its nest; but even if one was** found, and the young got away, there would be no certainty to have either the old birds or the young at the same place the following year. It cannot surely be supposed, by any reasonable person, that all the individuals of the rarer species of birds which occasionally come to us are killed or even seen; and if they are not, it seems to me that conclusions that they would breed here are arrived at **upon very insufficient** grounds. If the nest of a rare bird is found it should not be recorded from what has been told to any person only, but the fact should be capable of proof. What reason have we for supposing that the two Cranes noticed in Orkney or Shetland, and one of which was shot, were of opposite sexes? It appears to me that the chances against their settling there were a thousand to one. We cannot control the flight of migratory birds, or keep them one day longer than their appointed time. One of the Cranes having been shot would rather make it appear that they were not wild birds, or if wild, that they were out of condition and **lost**. I fancy it is no easy matter to shoot a wild Crane in good health.

The number of birds wanted for the legitimate study of Natural History, by almost any number of persons, will never be **missed. It is not killing a few birds** here and there that does **any** great harm. It is the improvement of the land which prevents Bustards, Cranes, Storks, and these kind of birds, migrating to us, or passing over us. Their day here (if they ever were common, which is doubtful, the name having in some instances changed by which birds were formerly known) is irretrievably past: even the beautiful Heron is decreasing in number as population increases and gets more dispersed. There is, unfortunately, no help for **it**.   C. M. A.

SAND GROUSE AND OTHER BIRDS.

(*Reprinted from " The Field" Newspaper*, **October** 3rd, 1863.)

SIR,—I have looked each week (more particularly since Partridge shooting commenced) in the expectation of seeing some notice of the occurrence of a young Sand Grouse having been obtained, which would have almost settled the question whether these birds had really visited us for the purpose of coming permanently amongst us, or might lead us to expect that they would return again another season; and I must add that I should have been pleased to see such a notice and to find that I was wrong in considering that the flight which did come had been lost, and had arrived here by mere chance; but, without having seen such notice, I cannot help thinking that, had they come to breed, their having been shot at would not have caused them to abandon their intention. Probably the whole of those which visited Europe would not be a greater number than might have been trapped in their native country in a single day. Should any of your readers have authentic information of any of them having paired, I hope he will be so good as to send it for insertion in *The Field*, where the notice will no doubt be duly appreciated. Of course, it can be of little importance to any one to keep back the truth, and what is interesting is the truth, and it only.

At the meeting of the British Association, recently held here, I understood a gentleman to give as the reason of the Woodcock becoming scarce in England that the eggs, being considered a delicacy by the Norwegians, were gathered there. Can any of your readers give us any information on this matter? In the first place, what proof have we that the Woodcocks which come to us do come from Norway? In the next, the shell of the egg is so very thin, it seems improbable that the inhabitants could pack them and send them to any distance to the towns; and they themselves probably eat all sorts of eggs they are fortunate enough to meet with; and the country over which Woodcocks breed is probably so thinly populated, and of such enormous extent, that it appears equally improbable that the eggs could be gathered in any quantities to do any damage to the species which could be observed.

The letter in *The Field* of the 15th August, headed "**The Little Bird**," is of great interest, could we understand the subject of it thoroughly; but on reading it attentively many things appear which require more knowledge to understand than many of us possess, or, at first sight, even think of. There are some statements which seem hardly what we might have expected; for instance, the observations about the Hoopoe, which you have already noticed. Are we to suppose that it kills birds? and, if it does, in what manner? and, if not, how does it get them to feed on?

It seems pretty clear from it that none of the Owls enumerated are game **destroyers, and,** perhaps, in consequence of nothing having been said against them, and so much said in their favour, they may be less persecuted in future. It also seems clear that **the** Jay and Magpie are not worse than the Rook, and therefore if the former birds are to be destroyed, why not destroy the latter also? It should not be forgotten that, although birds are devourers of insects, they not only take those which feed on vegetables, but they also destroy numbers of insects which are insect feeders themselves, and which are perhaps as great benefactors to man as the birds which feed on them are. Dragonflies and Ichneumons feed on insects; and even the Wasp seems to feed on Aphides, the larva of the Ichneumons, to such a supposed extent as in some **years to render the** perfect insect of some of **the** butterflies very scarce. The grub state of the Ladybird Beetle feeds on Aphides, and although when observed, as it is generally taken for a Caterpillar on the roses and other plants, **and** killed in consequence, it is really there looking for its food. This may be seen by taking one in your hand, and placing some Aphides near it; you may see it thrust its head into an Aphis, and suck it away till nothing remains but its skin.

Do not birds generally move about much more than is supposed? I have known many instances of the Kingfisher appearing in autumn at burn sides, where they certainly were not bred, or had not bred, and could not, as there were no bank sides suitable for them to breed in. The reason of their coming is not obvious; they come to particular places, and although they come to such places in autumn, sometimes each year, they do not remain long, but merely seem to come in passing. Was the Kingfisher ever a common bird in Northumberland? There probably are instances known of its having bred, but is it not possible that those which are to be found stuffed have been collected year after year for a number of years? and may they not have been shot during a temporary residence only, at the places where they were obtained? Many of the rivers in Northumberland seem scarcely suitable for it to breed near; they seem to be too rapid and stony. Are not the places where it is most frequently observed quiet, almost still, clear, deep waters, with occasional shallows, overhung with trees, and sometimes even at ponds? Where are the Kingfishers coming from, and whither going, which visit us in autumn? We generally suppose that birds migrate towards the south as winter approaches, but is the Kingfisher commonly met with further north? The Crested Grebe and the Eared Grebe are not uncommon on our coast, and the Little Grebe is common at some places in winter, but they do not appear to breed so far north as Northumberland.

I cannot help thinking that my argument must hold good to a certain extent, which is this, that many birds come to us totally irrespective of others of the same species having been previously here, whether killed, or protected, or not; and I come to the conclusion from the fact of birds, and even whole flocks of migratory birds, coming in autumn, which are often composed entirely of young birds hatched that year, no birds of greater age being observed either migrating before them or accompanying them. Certainly my observations relate principally to wading birds; but if it is so with them, why may it not be the same with others? I should state that the plumage of the young birds of that tribe varies so much from that of the old birds, that any mistake to a person acquainted with the species is not likely to occur. Many wading birds merely pay us migratory visits in passing and repassing in spring and autumn, and many of the species come only irregularly—some seasons passing without any being observed, some appearing only in autumn, others appearing at both seasons; probably almost all that do come are mere stragglers from the great flocks spreading themselves over the country in search of food in winter, and again returning to their breeding places the following spring.

The Ruff formerly bred at suitable **places in England abundantly**, but in consequence of its breeding places having **been** drained, it is now a very rare bird in most places in England in its summer plumage. The old birds probably leave us before moulting their wing feathers in September. I never saw one killed moulting its quills, and I therefore presume the old birds which breed in England leave so soon as the young can be left to themselves, the males perhaps soon after the females begin to sit; **and the old bird in its winter plumage is very much less frequently met with.** In September, some years since, flocks, composed exclusively of young birds of the year, used to come to Prestwick Carr, probably on their way from Sweden, and remained two or three weeks. Some years many more came than others. If some of these were killed, then, would that make it probable that a smaller number would come the next **year**? or, if none had been killed, would they remain one day longer? I think it unlikely. This bird occasionally bred there, and a nest **was** sometimes found **when no Ruff in full plumage had been** observed that year. I am convinced that birds move about much more than is **generally believed.** How soon the Garganey and Shoveller leave their breeding ground after the young can fly. I have known several pairs of the latter to breed, but so soon do **the** young take their departure, that not one has been shot in the neighbourhood between the breeding season and far into winter, **and** then the probability is that the birds met with have come from a distance. I never remember to have seen a Garganey in the London market amongst the quantities of wildfowl sent there during the autumn or winter, and therefore presume they retire to other countries so soon as the young are able to fly well; and it seems probable that most of even the commoner species of wild ducks which breed with us leave us in winter, and that their places are at that season taken by birds which come from a distance. I have seen flocks of young Teal migrating along the shore as early as July. Many migratory birds, particularly those which come in winter, appear to come in greater or less quantities according to the wetness or dryness of the season, the supply of food, and the severity or mildness of the season; perhaps not here only. Those species which come few in number and appear rare, are probably mere stragglers from the greater numbers which are passing eastward, and which are probably equally numerous as a species, but which come here, I think, totally irrespective of the number shot or caught here. When we take into consideration the number of such species as the Brentgoose and Wigeon, not only those which come to us, but those which spread themselves over other countries in winter, a few hundreds killed, more or less, really amounts to nothing, and no fewer will come the following year in consequence of their having been killed.

The Crossbill and the Chatterer seem to come most curiously, particularly the latter bird. Many years sometimes elapse between their visits. Can it be supposed that shooting a few when they do come has anything to do with the number that may come in after years? I think not. But at the same time I do not wish to have them killed when they do come, but should be only too glad to have an opportunity of seeing them every year, and in great numbers.

Can any one tell us any thing about the departure of the Corncrakes? They leave us in the same mysterious manner year after year, and we get no more knowledge about them. If man could control the flight of birds, either by protection or any other means, what capital sport might be had with this species. In summer, in many districts, every meadow holds a nest; yet, after their general autumnal migration—which must take place early and by night, or they would be observed, it is an uncommon occurrence rather than otherwise to meet with one, though sometimes one may be met with so late as December even.

Probably, were the feeding grounds of the birds not destroyed and curtailed, and they had space left them where they could sometimes retire to, we should still have abundance of them, in spite of all the shooting; but if you limit the ground suitable for them, comparatively speaking, to a small pond or a single marsh, and then persecute them, they must be driven away. Birds like to have room to choose particular places to feed at, and, if they cannot find such places here, their great powers of flight enable them to find them elsewhere.—C. M. A. (Newcastle-upon-Tyne.

ON WHAT ARE CALLED GENERA AND SPECIES.

.(*Reprinted from " The Field," Dec. 26th,* 1863.)

WHEN I first began to take an interest in Natural History I fancied the terms genus and species were easy to understand; but it is now long since I became acquainted with the many difficulties which arise before a conclusion can be arrived at. The former term appears to have no definite meaning: it **may either** comprise a whole division, as the fishes or the birds, **or a particular** tribe of each division, as the butterflies **or the hawks**; or it may be used in a very limited extent, and comprise only one or any greater number of what **are** called **species**, according to the fancy of any individual. It is certainly useful to have what are called genera, by which **nearly allied species can be classed together**; but when so many genera **are recognised as at present, you may almost** as well separate **each species into a genus at once.** The term species is much more difficult **to deal with. Before endeavouring to determine what constitutes a species, would it not be well to** expunge those animals which **have been in a state of domestication** for a great length of time? and for this reason—that it is at the present day quite impossible to trace back what they originally sprang from, and also to trace by what, **if** by any means, such animals may have got distributed over the various portions of the globe. In some instances, climate, food, and (having escaped) running wild almost in a state of nature, and other causes, probably have effected such changes, that animals descended from the same original stock are now by many persons considered different species; and on one being brought in contact with a different looking animal in another part of the world, and breeding, it may lead persons to suppose they have got a hybrid which is capable of reproducing, when in reality the produce is not a hybrid at all. Nature appears to have taken especial care to prevent the mixture of species, and wisely: if it were not so, instead of having the beautifully-defined forms we have, we would have the world full of I may say monsters. How species are or were created is a matter quite beyond man's understanding; we cannot tell whether species are being created at the present time, though we have such abundant proof of their dying out, and having died out. There does not appear to me to be any thing more to be wondered at in the creation of a species at the present time (should the Creator of all things deem it right) than there was at the beginning of the world, but by what means man will probably never know. I cannot believe man can make a species, which he would be doing if he could obtain a hybrid between two distinct species, and continue the breed of such hybrids constant in shape and colour. If nature approved of hybrids they would be much more frequently met with in a wild state. Undoubtedly they do occur in that state, but much more frequently amongst those animals which are either domesticated or semi-domesticated, and in cases where the parents are deprived of the opportunity of associating with others of their own species. When hybrids do occur in a wild state, does nature not prevent their doing any mischief and interfering further with her regular laws? As a general rule, would it not be well to regard those as species the parents being alike, the offspring similar, and whose young reproduce? Almost all rules have exceptions, but nature has generally given certain colours to a species in its wild state at the same season of the year, and at the same period of the age. I see in *The Field* notice of what are called hybrids between the Japanese Pheasant and the Common Pheasant. A question arises—Are these separate species? Should they be distinct species, and those bred really be hybrids, it is more than likely they will prove barren. Is it not probable that those Pheasants, and also the Ring-neck and the Bohemian, are mere varieties, caused by food or climate, or other cause? in which case the young are not hybrids, any more than the varieties we find amongst domestic poultry. Another question also arises, which is—Can any substantial good arise from producing hybrids? Supposing their flesh is superior to either of the parents, is it worth the trouble to get the produce over and over and over again, as are not the hybrids as a rule generally barren? I mean those bred between really distinct species. One often hears a person remark amongst domestic poultry that some of the hens being spotted, similar to the black marks on the breast of a cock Pheasant, that they have a cross of Pheasant in them. Now, instances of a cross between the Pheasant and the Domestic Fowl are very rarely met with. I have seen the cross, but they were obtained by having the Pheasants confined, and the mules are as much like Pheasants as Cocks, and they vary in colour. One sees other undoubted hybrids amongst birds, but it appears to me anything but desirable to have such mongrels, except as curiosities, as they are not so handsome as the parents; and are they not generally barren? I confess that I have a great dislike to hybrids of all sorts. Nothing can be more interesting and beautiful to behold and study than the various species of wildfowl in a domesticated state, but so soon as you mix hybrids with your collection the interest of the admirer of nature ceases, and it only confuses a student who may be desirous to get a knowledge of species seeing these generally useless hybrids mixed with nature's beautiful species. The various changes of plumage that she has given to the various species at the various seasons are quite puzzling enough. No one can imagine why some species of birds which inhabit Europe and North America appear similar, while in others, though closely allied, there is constant and marked difference. It is probably most advisable to consider at present these differences where they are constant, however trifling they may be, to denote a species, particularly when they inhabit a different portion of the world, and there is little chance of their ever meeting with each other. Some of the domestic animals have altered slightly by domestication, except in colour, as the Duck; but then probably wild Ducks have been often caught, or hatched and reared, and thus the natural colour and **form have** been infused frequently. In the Common Goose some of them are so like the original species in all apparent respects, that I cannot fancy why it should ever have been supposed they have any trace of either the White-fronted or any other species. It always strikes me as curious why the various species, taking all known into account, do not differ from one another in a more equal degree. We find two very nearly allied, then we have to look some distance; we then find perhaps several, then again there is a wide gap in the chain, and so on. I do not wish to be understood that we should expect to find one species so like another that a regular arrangement could be made; that is quite impossible, as one species sometimes resembles another in one respect, while in others it resembles other species again further off; but, for instance, if the American Wigeon and Teal are different species from the European, where is the next species on either side of either of them connecting either of them as closely to any other species? C. M. A.

ON HYBRIDS.

(Reprinted from " The Field" Newspaper, Feb. 20th, 1864.)

In the notes from Aldeburgh two ducks are mentioned as having been killed, which the writer calls "Cross-bred Wild Ducks," and says they seem to partake of the markings of the Sheldrake and Common Duck and Mallard; but he does not give a sufficient description to enable any one to form an opinion as to what the birds really may be, and it certainly is worth while ascertaining what they are. The fact of two hybrids being together, particularly if they are of the cross the writer supposes, suggests the probability of their having escaped from captivity, the habits of the two species being in the wild state so different, and no instances of a similar hybrid have been observed that I am acquainted with. As uncertainty exists as to what should be called hybrids, if **crosses** between varieties are, as well as crosses between what are at present considered well-defined species, some distinguishing name is required to mark a difference which appears to be considerable and definite. As a general rule, in a **state** of nature either one or the other seldom occurs. Many persons fancy there is as little difference between the Pintail and Wigeon, or Gadwall, as there is between a small Decoy Duck, a what is called a Buenos Ayres Duck, or an Aylesbury, or any **other** domesticated variety of *Anas boschas*. What difference there is, is of a different kind altogether, and consequently, the crosses between the former class and those between the latter **are** also totally different; the former cases occur in the wild and natural state, the latter cannot, as these varieties have undoubtedly originated through domestication only. From the very few hy**brids** I have seen killed in a wild or natural state, any one possessing a knowledge of the species may at once detect the parents **from** which they descended; but we have no examples that I have ever seen where the two species became again more mixed and less apparent; and I think it questionable if such an instance in the wild or natural state ever existed. Hybrids amongst birds in the really wild and natural state, appear to occur in a very limited number of genera and even of species in these genera, and much less often than is generally supposed. It might throw some light upon the subject if some of your correspondents would inform your readers of any instances which may have come under their observation, and at the same time say where the specimens can be seen. With the exception of those between the Pheasant and Black Grouse, and the latter bird and the Capercailie (the former cross cannot be considered a cross between birds in a state of nature) (but even in these cases no further cross appears to occur,) I have only seen instances amongst the Ducks, and it seems probable if all those known were added together, one would not occur annually amongst the thousands of wildfowl which are taken. In the wild state crosses between extraordinary or occasional varieties, as albinos, of the same species really seldom occur, the instances of such varieties are so very limited in number; and, as a general rule, the variations amongst species are so trifling, that it is difficult to describe how they vary, each individual probably being a shade lighter or darker, larger or smaller, or more or less spotted or marked. True it is, we find sometimes a wild white or pied bird, but it would appear that nature does not like these, as they do not apparently increase in number; and probably if pains were not taken to keep the particular varieties of species amongst birds produced by domestication separate, we would find that the varieties would gradually become assimilated, and probably we should in the course of time get a variety from the whole which would be best adapted for the climate and circumstances, and probably it would much resemble the original form of the species from which the several varieties descended. Many of our native species refuse to breed in confinement: some do breed, but recent domestication does not seem to alter their general appearance, and probably, in spite of all experiments made with them, except crossing them with other varieties, they will continue as they are. In a notice which I observed lately in *The Field*, I saw it stated that Dr. Fleming considered the Wild Duck (*Anas boschas*) as we have it, the parent stock of all Ducks. Is it to be inferred that merely the domesticated varieties of *Anas boschas* are intended to be included; or are the various species in England, and the foreigners also, to be included? If the Pintail, Gadwall, and Shoveller are included, why not include the curious Shoveller from New Zealand and elsewhere, and the Bahama Pintail? and from that again our Long-tailed Duck also? and, if so, where are you to stop? I should like to know if a bird exactly resembling the present form of the old *Colchicus* (Pheasant), which probably flourished in England before *Torquatus* was turned out, is to be found in a natural and wild state in any part of the world at the present time, so as to ascertain whether time and circumstances may not have moulded it into the form best adapted in its now naturalised state in the various countries in which it is now found, and which might account for the various colours of the varieties of this Pheasant found at the present time in China, Japan, and elsewhere, and which have probably all sprung from the same original stock, and through domestication. I believe the ring on the neck occurs in many Pheasants which have nothing whatever to do with *Torquatus* recently turned out. The peculiarities respecting the breeding of pied and white Pheasants do not show that these varieties are any thing more than varieties originated by domestication; and although they may be by some persons considered pretty, their white feet and piebald appearance remind one that they are not wild birds, and too much assimilate them to the inhabitants of the poultry yard. I should like also to know if there are at the present time in a naturally wild state, in any country, either the Muscovy Duck, from which our domesticated birds are descended, or what is called *Cygnus anseroides*. I think I am right when I say I remember a pair of Wild Swans, in the year 1840 or 1841, hatching and rearing a brood in the Zoological Gardens. On going there some time afterwards, I missed the greater part of this most beautiful family. On inquiry I was informed, whether correctly or not I know not, that it was wished to get a breed between the tame Swan and the wild one, and that all, excepting one parent, had been sold; and I can recollect my feeling of disappointment, and my wonder that any one could suppose, or presume to suppose, that the extreme beauty of *Cygnus ferus* could be added to by such a mixture.

C. M. A.

ON SUPPOSED HYBRID GROUSE.

(Reprinted from "The Field," 1864.

I READ in *The Field*, a few weeks ago, of the capture of a bird, supposed to be a hybrid between the Black Grouse and Red Grouse. I possess a bird which some persons have said is such a hybrid, but it undoubtedly is no such thing. It is a Black Grouse (probably a female, having assumed, from some unknown cause, some of the peculiarities of the male, or it may be a bird which has always had a similar plumage); it is greyish-brown, darker than ordinary grey hens, with blue reflections on the neck, and having the tail curved outwards like the male, and the under tail-coverts white. I have seen two or three similar birds. Now if these birds were hybrids they were males, but they want the large scarlet process above the eyes which is conspicuous in the male bird of each species; and whilst all the characters of the Black Grouse are present, those of the Red Grouse are absent. Is the bird referred to similar to these? I do not say it is impossible that such a hybrid might occur, after seeing the singular varieties of Russian Black Game, which are varied with white, and which seem to be hybrids between that species and one of the Ptarmigans. The only object in questioning the accuracy of the fact of such a hybrid having occurred is the desirability of not extending information which may not be correct. I have certainly never seen such a hybrid, nor heard of an authentic instance of one amongst the thousands of Grouse which one sees each season.—C. M. A. (Newcastle-upon-Tyne).

HYBRID GROUSE.

ONE of your correspondents (January 14th, 1865,) writing on this subject, begins by stating that the "Rype" (which is what we call the Willow Grouse) and our Red Grouse are the same species. How is this **ascertained? Some time** ago, **I** think, **I read** in **your paper that some one had tried** to introduce the Red Grouse **into either Sweden or Norway, and** I have been expecting **to see** a notice of the failure or success of the experiment, as I expected **to be** able to come to some conclusion from the results. Now, it would be a most interesting fact to ascertain if these local varieties or species, or whatever they may be, will breed together; **and,** as communication is now easy between the two countries, could **and** will not some extensive landed proprietor get over some Willow Grouse and turn them adrift among the Red Grouse, **and** see if they **will** breed with them, and produce fertile young? **This** would probably satisfactorily settle **the** question, and, if they did breed, it would be quite curious to observe whether the white wings of the Willow Grouse would continue, and the birds get white in winter, or whether our climate would cause the original local colour of the country (supposing the species were the same) to prevail. If the Willow Grouse **were** brought over **and bred** by themselves, it appears to me the species should **be** considered distinct. I hate mongrels, and would never advocate the mixture of different species, as I do not believe any good results will ever be thus produced in animal life; but proving whether these **two** Grouse are the same species would go far towards enabling naturalists to give an opinion on other species which **may** be distinct, or which may be only local varieties, but yet constant, from **the** fact of their never having the opportunity **of intermingling** with each other. **I am no** believer in the deterioration or improvement of species in a purely natural state. The first bird "Scrutator" writes about, he says, was the weight, size, and shape of a full-grown Black Cock. Now, if he were a hybrid between the Red and Black Grouse, he ought to have been about the size of a Grey Hen: the legs and feet of both species are hairy. He does not mention the toes, which are hairy

in one species and not in the other; and the nails are very different. It sometimes happens that a male bird assumes some of the plumage like that of the female: Is his bird such an one? My bird, according to his opinion, should be a female hybrid, as it is of the **size of a** Grey Hen only, but with a Black Cock's tail. Now, if a female, why does the tail at all resemble that of a male Black Grouse? it should be intermediate between the tail of a Grey Hen and a Hen **Grouse**. It strikes me the only feature **in** these birds resembling the **Red Grouse** is the colour, **but even** it much more resembles that **of the Grey Hen;** the birds are much freckled with grey, **and** have not much red, like an old male Grouse; but even if they had much red, **many** Grey Hens have a considerable portion of rich reddish brown mixed with their plumage. Are the hybrids between the Capercailie and Black Grouse not all males, or nearly so? I do not think I remember seeing **a** female. These hybrids appear to have neither the tail of one species **nor** the other, but a modified one, and the individuals partake evenly of the peculiarities of **each** species in size as well as colour. "Scrutator" will, **I hope,** pardon my **still not** being convinced. I only wish **to arrive at** what is correct.
C. M. A.

SOME OBSERVATIONS ON THE BREEDING PLACE AND EGG OF THE KNOT, TRINGA CANUTUS.

(Reprinted from the Transactions of the Natural History Society of Northumberland and Durham for 1877.)

I HAVE wondered ever since the last Arctic expedition started why this species was especially mentioned to be looked after, as I think there are many others of which almost as little is known, which breed in the Polar regions.

I find amongst **my** ornithological memoranda, an extract I made in 1839, at Wallington, from Captain Sabine's account of **birds** found at Melville Island and the North Georgian Islands during **Parry's** voyage, which is as follows:—

"*Sanderling*. Breeding abundantly on the North Georgian Islands.

Golden Plover. Very common, North Georgian Islands.

Ring Dotterel. Common, both places.

Turnstone. Very common.

Dunlin. Rare.

Knot. Breeding in great abundance on the North Georgian Islands.

Purple Sandpiper. Breeding all along the coast of Davis' Straits and Baffin's Bay, but not met with in the Polar Sea.

Grey Phalarope. Breeding abundantly on the North Georgian Islands;" and amongst several other species is mentioned that lovely bird the Fork-tailed Gull as occurring on Prince Regent Island, and on three small Islands in Baffin's Bay, in lat. 75·5°.

Now this seems to me a very satisfactory proof of the Knot's breeding ground being ascertained in America at least.

I am not aware whether any eggs of the Knot have been brought to England, but I should think it very probable they may have been. Some thirty years ago I was staying at a house in this county, when I was shown a collection of eggs, and there was one amongst them named Knot. The original collector had died, but I enquired about the egg, remarking on its rarity, if authentic. My kind correspondent (also long since dead), a few days afterwards, wrote to me as follows respecting it.

"The Knot's egg," he says, "he had from a Captain Howard some years ago. He was a great collector of eggs from all parts, and he believes it to be certainly correct, I therefore send you the egg, hoping you will put it into your collection if you think the information worthy of belief." The person referred to was the keeper, and I believe was a clever man, and was much interested in birds and eggs, and had helped to form the collection. This egg agrees pretty well with Yarrell's description of what the egg is said to be like. It is about the size of a Reeve's egg, but is rather broader and not so pyriform, the ground colour being a very rich pale brown with a pinkish tinge in it, and the spots are rich brown, with some inclining to lavender colour. Now from the shape of the bird one might expect it to lay an egg shaped like the Woodcock's egg. I admit shape goes for very little, as I have Dunlins, Common Sandpipers, Curlews, and even Snipes' eggs, all of which are generally very typically pyriform, as broad and round as Woodcocks, but I think I never saw a pyriform Woodcock's egg. In the days when this collection was made, and previously, collections of such like things were more often to be found in country houses than they are at the present time, but the opportunities of getting from home were **not then so** numerous and consequently public collections were **not so** easily seen, and those who had any taste for such things collected for themselves, and I think generally, in consequence, the collectors acquired more information than the generality of those you now meet with care to possess.

I would remark that it is possible the Golden Plover and Ring Dotterel and the Dunlin mentioned, may not be those of Europe; it is very difficult to say whether the American forms of these birds are the same species as the European.—CHARLES MURRAY ADAMSON.

THE KNOT.

HAVING copies of Captain Sabine's memoir of the Birds of Greenland, and also of Richardson's Fauna Boreali Americana, I find in the former this bird is mentioned as having been killed on Hare Island in June; and in the latter it is stated it breeds on Melville Peninsula, and in other parts of Arctic America, and also in Hudson's Bay, down to the 55th parallel, and that it lays four eggs on a tuft of withered grass, which are, according to Mr. Hutchins, of a dun colour, fully marked with reddish spots.

Faber mentions the Knot as found by him here and there in Iceland, particularly on the southern sides of that Island. Its arrival is late, about the last days of May, and it has then assumed the blood-red under plumage of the summer. At that time it is to be seen on the shores in company with the Turnstone, and, he adds, fully equals this species in its excessive wariness and difficulty of approach, and in summer that it disappears from the vallies and "probably" betakes itself to the mountains to breed. They reappear clothed in their winter dress in the beginning of September, and in about the middle of that month take their departure from Iceland.

Faber seems to have mistaken the young birds **for** the old birds changed to winter dress, and which there also, **as** with us, appear to keep in separate flocks from the old birds. But what really does become of the greater portion of the old birds, until they come to our shores in late autumn in their complete winter dress?

There can be little doubt but that the breeding grounds of the Knot are very extensive, and probably hereafter a colony may be found at Spitzbergen from whence those which visit us may come. These birds seem similar in America to those we get, but the American Dunlins, so far as I have seen, are considerably larger than ours, so much so, as I think to constitute a race, if **not** a species.

When in London in June, 1878, I saw three of these interest**ing** birds in their summer plumage, which they had acquired in **the** Fish House in the Zoological Gardens, and I was delighted **with** them. There were also a Turnstone which had acquired its summer plumage, two Avocets and two Grey Plovers, **one a bird** of last year, which had nearly acquired its winter plumage, **not** having any black on the breast, and having still some of the spotted feathers of the young bird amongst the tertials, the other **an** old bird, about half changed to its summer plumage, that is, irregularly spotted with black on the breast. These Grey Plovers readily called attention to them by frequently uttering their well known call note. I also saw in the British Museum the three young Knots in the downy state, which had been brought home by Captain Fielden of the Arctic expedition. On looking at a label on the case I saw the date June 4th. Now, I felt sure this was a mistake, as the grounds on which this species breed could not be ready for them by that time. I find, on looking into the appendix to Captain Nare's book, I was right, as it was not until the 30th July these young birds were obtained; but I also must add, that I consider Captain Fielden wrong when he says the Knots had acquired their autumnal grey plumage by August 25th, unless he means the young birds' first plumage, as we have the old birds sometimes returned to the Northumberland coast by the middle of August, which have still their faded red plumage, and have not cast a single feather. The grey birds he met with in August were I think without doubt the young birds of the year in the very same state I have frequently shot them towards the end of August on our own coast. I was almost sure Knots had not laid eggs so early as June 4th; I have had them killed on the south-east coast till the beginning of that month on their way to breed.

It seems distance within reasonable bounds is nothing to some kinds of birds. Knots, and some other kinds, leave here only in the end of May, to breed in the Arctic regions, and by the middle of August the young have flown here, the old birds having bred and reared young capable of flying the distance in the short space of ten or eleven weeks.

The six birds mentioned by Captain Fielden, as seen by Dr. Coppinger on the 28th July, and of which an old male only was procured, might possibly be an early family, the four young and two parents, but it seems strange they should have been so wild if such was the case. This species when young and in small flocks, and apparently unaccompanied by old birds, is often tame and easily approached, but at other times I have seen them, even on their first arrival late in August and early in September, when in large flocks so wild, that they would not let you approach within half a mile of them.

On the 6th September, 1844, I got a young Knot which had flown against Skinburness Lighthouse, which was of wood standing in the sea; the beak was broken in the middle as if done with a pair of pincers: and one young bird I shot on the 10th September, in the same year, had at that early time commenced to moult to its winter plumage, the back and breast having imperfect feathers all over. I was struck with the difference in the habit of this bird in different situations. On the Northumberland coast they are often found amongst the rocks as well as on the muds, and very tame. At Skinburness when **found they** were on the grass and wild: they were driven up by the **high** tides and were sitting upon small hillocks of grass surrounded by the water in similar places as were young Ruffs, **so** that what is mentioned about them having been caught with Ruffs in former times is quite probable. Previous to seeing them in such places **I** thought this a mistake.

North Jesmond,
Newcastle-upon-Tyne,
August 1st, **1878.**

ON THE HABITS OF THE KNOT AND OTHER SHORE BIRDS AS OBSERVED IN NORTHUMBERLAND.

(*Reprinted from " The Field"* **Newspaper.**)

THE Knot is well known to visit our **southern and** eastern coasts each spring and autumn, and **in May it appears in** its **red** summer **plumage,** or is then acquiring it. **On** the Northumberland **coast I am** not aware **of** a single red bird ever having been killed in spring when the species is migrating north, **and** from this we might suppose that the flocks visiting our more southern shores keep further to **the eastward** in their northern **migration,** and that the numbers **which** visit us in autumn are **European or** Asian, **and** not **American.** Singularly enough, the **old** birds occasionally come to the Northumberland coast after breeding **without having cast a** feather. The earliest autumn arrival I ever knew I **shot** on July 19, 1854. This bird had bred as the bare places on the breast clearly showed. I have some old birds killed early in August, and have seen several others, more worn and faded in feather than the one before mentioned. These birds are not accompanied by the young, and they have not commenced to moult. By the last week in August the young come often in large flocks by themselves. It may be the few old birds which arrive so early have lost their eggs or young, and, it being too late to breed again in regions where the summer is so short, have come straight away. In autumn the old birds (except the few which come so early) do not seem to arrive till they have completed their winter plumage by moulting, and the reason of this probably is that in September they would be moulting their quills, and unable to take the long flights requisite to bring them so great a distance. Although the young birds arrive on our coasts so early, there has nevertheless, been time for them to get here. The old birds apparently arrive at their breeding stations in the middle of June, and from the time they lay their eggs till the young are fully able to fly would be from nine to ten weeks, which gives them time to reach us by the end of August. But where do the old birds linger during the moulting time, both those which come away early, and those which come later? I have never met with one moulting its quills. I have an old female knot which I bought in Leadenhall market in May, as dark in colour as the female Grey Phalarope is in summer, and darker than any male knot which I have seen. Now, as regards the young which arrive in August and September, the males are rich buff in colour, whilst the females are very much greyer. The change in colour of the old birds between the time of their leaving us and the date of their return is very apparent. The red breast fades to sand colour; the pink spots on the back become nearly white, and the grey edges wear off the back feathers, which become black. The young birds somewhat resemble the old birds in summer, but are much paler in colour, and gradually both old and young alter much as the season advances, and by the year's end they get very nearly alike; but the moult in the young birds to winter plumage is not nearly so complete as in the old birds, which cast every feather in September, the young birds always retaining the primaries and secondaries, and many other wing feathers, till the general moult the following autumn. It is rather singular that the young Sanderling, Pigmy Curlew,

Grey Plover, Knot, Bar-tailed or Common **Godwit, and Little Stint,** should all arrive **here** about the same **time, some of these species being found in America as well as Europe, and others not.** The Knot and Common **Godwit (the former common to both continents, the latter not) seem to come almost in the same flocks.** This year, whilst **at the** seaside on **Sept. 6, after a morning of heavy rain and strong** northerly wind, the afternoon being sunny, **I walked out on the** sand, where I saw a large flock of Godwits, probably just arrived. They looked quite buff in the sunshine, and **let me walk** pretty well up to them, and I shot eight, and one Knot, all young birds—an interesting **lot,** as it enabled me to make comparison of the **individuals, and observe the difference** in the size of the males and females. **Some of the latter exceeded by 3in. in extent of wing some of the males.** I find I have observed **the beak of this species to vary in** length from 2¼in. to 4½in. They have a pretty cry, somewhat resembling some **of the** notes of the clarionet. Both these species when they arrive are excellent eating.

But the question **is, whence come they? From the abundance** of both these species, their breeding grounds must be very extensive. I think it is unlikely that such birds would travel over a continent entirely of land for any great distance. More probably their course **is along shore;** but how difficult to trace. In the young of the Common Godwit the males and females are equally rich in colour, whilst **in the old** birds in summer plumage the male **is very much darker than the female, which** often shows very **little red at all. The Common Godwits,** Grey Plovers, and Knots which **visit, in their spring migration, the more** southern and eastern counties of England, and do not come on to this coast at that season, may possibly pass eastward and travel over the lakes and seas and narrow parts of **land** until they reach the White Sea, **and pass on till they arrive at** a sufficiently high latitude in **Asia to breed, as there seems no** land in Europe parallel with the **North Georgian Islands were** Capt. Sabine states the Knot was found breeding abundantly; or they may go round by the North **Cape; but we do not** often hear of their being met with on the journey. The great migration probably passes further eastward than Great Britain, as the few which alight on the English coasts must be but a very small portion of the migrating host. Where the tides run out a long way and leave a great extent of sand and mud, it is uncertain work following these birds at low water; but at times, at particular places, during spring tides, at high water, numbers may be shot. One day under such circumstances, on Sept. 19, with a northerly wind and high tides, I got many Knots and Godwits and a couple of Grey Plovers; the following day at the same spot, with a southerly wind and a tide not coming within one hundred yards of that of the preceding day, I did not get a bird, none being driven up, in consequence of such an extent of mud being left uncovered by the water even at a spring tide.

The Sanderling, like the Knot, seems to be as abundant in America as in Europe, and its migrations seem very similar. Large flocks, however, in going northward, appear on the west and more northerly coasts in May, and these may be on their way to America. The Grey Plover, also, like the Knot, seems to be equally distributed in America as in Europe and Asia, and its migrations extend probably farther south than even those of the Knot; but little is apparently known about it. The earliest bird I have seen arrived from its breeding grounds is a female, killed Aug. 20, 1862, the only one with a black breast shot on our coast that I know of. The young birds arrive about Sep. 16; they are conspicuous by their pretty spotted plumage, showing the white back feathers when they rise, and often uttering their loud cry of "kle-wee." They seldom appear in large flocks, being generally seen in straggling parties of about eight, but oftener still in smaller numbers. The young Grey Plover when it first arrives and before it gets faded, when first killed and if kept clean, is an exquisite bird to look at, the lovely freckled back being a mixture of olive, white, and yellow, recalling the readily conceived place of its infancy. One can imagine the lichen-covered stony ground, with the brood lying well concealed by their similarity to the surroundings, and no doubt can be entertained that the wonderful arrangement of the colouring of its plumage is designed by nature to protect it from observation. How different the birds look when first killed to what they do when skinned or set **up**, under the most favourable circumstances. One might also **inquire** why the Phalaropes do not return from their breeding **grounds to** us in the same way as the other **northern** birds **mentioned.** Where do they stay on the road? and **why do the few** come which pay us a passing visit in early winter (that is, up to January only)? If going southward, why do not more pass? The red-necked species seem to leave their breeding-grounds (if we may judge from the few which breed in the Scottish isles) very early; but whence do they go to spend the winter, and how do they arrive at their breeding places without being more often observed when leaving and returning to them.

C. M. ADAMSON.

Newcastle-upon-Tyne, **December,** 1876.

SIR,—I am glad to see that, on the whole, Mr. Booth agrees with me. He remarks that the old and young Knot appear to be alike when once they assume the winter dress. This I admit; but the young do not acquire a complete winter dress the first year by moulting; the old birds do, completely, and suddenly, by the autumnal moult. The young acquire their winter dress gradually **and** only partially by moulting. Up to the year's end you gen**erally** find **some** of the tertials still edged, and the primaries and secondaries are not **cast** till the general autumnal moult the **following year**; and the wing coverts are those of the young bird's **first** plumage, which fade and wear into a winter plumage, but which can be distinguished from those of the mature bird on close examination. Indeed, it is often quite possible to trace the young birds by the wing coverts up to the time of their acquiring their summer dress in the spring, these feathers often getting by that time worn nearly threadbare—they being at first the more slender feathers of **the young bird**; and, besides this, they have been used by the bird from the time it was first feathered till the following spring, instead of having been renewed by moulting, as in the old birds, in September. We frequently find the Knot and Sanderling in spring having two or three wing coverts in the bright summer dress amongst the worn feathers, and sometimes a centre tail feather only in the same state. These, I think, are only makeshifts, to enable the bird to put off till the general moult in autumn, the bird requiring some new feathers, and, it being in the seasonable state, they come coloured; but, even these are probably cast again at the **autumnal moult. After these birds have gone through the autumnal moult the year after being hatched, I do not believe it possible to determine their age. The Grey Phalarope perhaps illustrates my remarks better than the Knot, as the plumage is more decided between the old and young birds in autumn.**

During the many years I have paid attention to birds, I never but once saw a mature **Grey Phalarope in winter dress** procured **in this county, and it is now in my possession. It** was killed **in** October **and had** then **completed its moult to the** sombre **grey winter dress. All the others I** have seen were young birds **of** the year, and undergoing their gradual change **to** winter dress. **I have them** in the young plumage, scarcely having any winter **(that is,** grey) feathers on the back at all, and showing darkish **reddish or** brownish colour on the neck, resembling the old bird in summer; and I have them **nearly** grey on **the** back, further **advanced** towards the winter dress; but all these are young birds, **and** have **the** dark tertials edged **with** light colour **of** the young **bird, and** these they have **up to the** year's **end, the** latest time we ever see them here. **The** plumage of the **old** birds is much more dense, and the wing coverts are grey, and not dark, as in the young birds of the **year. The** Grey Phalarope in its young and winter dress much reminds me of the black-headed, and little Gulls at the same season, as there is a similar difference **between the** mature and young birds in autumn and early winter. That **the** autumnal **moult** in old birds is rapid and complete, is certain, as I have shot mature Common Godwits and Sanderlings, completely through the moult, by the middle of October.

Though the winter plumage of the young Knot resembles the old bird, the winter plumage of the young Common Godwit does not appear to do so. In October the young birds begin to throw out a new set of feathers, somewhat resembling the first feathers in colour, but more marbled, somewhat resembling the Grey Plover in winter; and in this state they are sometimes seen in the following May, without any apparent approach to summer plumage. By December, all the back and breast feathers of the young birds correspond; the tertials remain unchanged, and the light-coloured spots on these feathers wear out, causing them to have a jagged appearance. This is sometimes the case with birds that have acquired the summer plumage the following spring, showing them to be young birds of the preceding year. In the winter plumage of the old birds the tertials are plain grey. I have a very singular variety shot in December, with a marbled back, the tertials much worn, showing it to be a bird of that year, with the breast quite dark, almost brown, but not red; and I have a male Common Godwit, killed in spring, which I got in Leadenhall Market, with many others, more than thirty years ago.— It is in very complete summer plumage, and has some of the feathers between the scapulars broadly edged with red, and not spotted with that colour as they ordinarily are.

It is not easy to account for the arrow-head shaped spots on white feathers, often seen amongst the red breast feathers of some male Godwits in summer plumage. These spots appear to form part of the female's ordinary summer plumage; but, generally speaking, they are not found in the male in the young plumage, or in winter, or summer. The mature birds in winter get so grey as almost to lose the bars on the tail centre feathers entirely, and they are hardly traceable in some males I have in summer plumage. In well-marked Grey Plovers also in full summer plumage, the bars on the centre tail feathers often become obliterated.

I have a young Grey Plover of the year, killed in December, considerably moulted to its winter plumage, but the new feathers have come marbled with grey, black, and more yellow than in young bird's first plumage. The yellow is marbled, and not spotted, as in the young bird's first plumage.

It is difficult to reduce to writing what one wishes to point out concisely. I have one of the larger young Knots (which I conclude are females) I shot at Skinburness Sept. 10, **1844**, edged all over the back with grey, and again with white, without a trace of buff colour. Then, **again,** I have the smaller bird, which I conclude is the male, quite buff,* much more strongly marked; even the centres of the **scapulars are** marked with **buff and darker** grey markings than the edgings. **I take these** as my types, **on** shooting several, I have placed the larger and smaller separately, and I generally found the larger the paler birds. I admit I never dissected these young birds, but I have dissected the old birds in spring, when it is easy to determine the sex; but to do this with **the** young birds would require a much more scientific anatomist than I pretend to be, as it is no easy matter to do so.

Audubon mentions his surprise at the difference in the size **of** these birds he killed from the same flocks. Should Mr. Booth come into this neighbourhood, I will gladly show him the birds from which I have drawn my conclusions. He mentions the appearance of young Knots so early as the 8th of July.† This is really most interesting, as it proves the bird to be an early breeder somewhere or other. We naturally supposed all **Arctic birds** were late breeders; from such an early appearance, it is impossible they can be the young of the birds only going north in the middle of May, so that there must be **an** earlier migration. In September, 1866, I wrote in *The Field* about the Godwits, Knots, and Grey Plovers **which had wintered in** this northern district, **leaving** so early **in the** spring, before acquiring their summer

---

* I have very considerable doubt whether some of these very rich buff birds have not been mistaken for the American Buff-breasted Sandpiper: few persons not very well acquainted with the species would take them for Knots.

† Could these be birds of the former year?

plumage, and so late in the spring appearing again on the south-eastern shores; but no light appears to have been thrown on this singular apparent double migration. I have seen Sanderlings till June 4 on the Solway shores, and **just before that** time they **are** often flying in flocks with pale-coloured **Dunlins; whilst the** rich-coloured **Dunlins** have eggs at a very **short distance.** These Dunlins **have black breasts,** but the backs show little **red,** though the birds **are in** summer plumage. The young Sanderlings are most common on this coast in late August and early September; the flocks after this appear to continue their migration.

Newcastle-on-Tyne, Jan. 1, 1877,   C. M. ADAMSON.

Since this was written I have reason to believe some Knots and Common Godwits do not acquire any red or breeding plumage the summer after being hatched, **but** occasionally remain on or visit our shores during the summer months with white breasts.

THE SANDERLING.

(Reprinted from " The Field" Newspaper, August 23rd, 1862.)

I see in *The Field* a notice of the occurrence of a bird called a Sanderling near Sutton Park, and some remarks on its habits, etc. Without questioning the power of the bird to either swim or dive in cases of emergency, I think some doubt exists whether some other bird has not been mistaken for it. The Sanderling is, I believe, one of the birds whose breeding-places in Europe are yet unknown to us. It is absent from **our** island perhaps as short a time as any of our migratory birds. I have noticed their first appearance on the Northumberland coast as early as the 31st July. Those which come so early are old birds, in summer plumage much faded, having the bare places **on** their breasts showing they had bred that season, and **commencing** to get some of the ash-coloured feathers of winter **on their backs.** Some small flocks of old birds **arrive** during August; **they do** not seem to remain long, as old **birds are seldom seen a little later** in the season. By September **the young birds arrive in flocks;** by October some old birds are also on the coast, which have then got through their autumnal moult, which is the only complete moult during the year, and are in what is called the winter plumage. Some are found on the shores during the whole winter. I once observed a young bird in August with some of the down not then worn from the tips of some of its back feathers.* By April, most of those which have wintered with us, leave. Towards the latter part of May, and till the first week in June, flocks appear on the coast. Most of the birds are then in their summer plumage. On the shores of the Solway, at that season, they are sometimes in large flocks; they remain a short time; after this they all disappear to breed. I never knew an instance of the Sanderling being seen away from the sea coast; it is generally observed close to the edge of the sea, wading amongst the sand, or muddy sand as it is disturbed by the tide, picking up its food. It is a most active bird on the wing, and generally wary when in large flocks, wheeling and flying about the sands; and, especially when in summer plumage, it is pleasing to see the pure white of the under-parts contrasting so well with the rich buff colour of the upper plumage. The young birds of the year can be distinguished from the old birds when they have got their grey plumage on the back by the wing coverts, and the long tertials, the young birds seldom changing them before the general moult the following autumn. It is difficult to account for the appearance of the old birds so early in the season. I have sometimes thought it probable that the few which do come so early may have lost their eggs or young by accident, and migrated earlier than they otherwise would have done in consequence. The old birds observed so early have not commenced to moult their quills; and I think they do not remain on this coast during the time they do moult their quills, as it is not usual to meet with them doing so. I think the greater portion of the old birds which visit us during winter do not arrive on our coast until after the autumnal moult

* In Captain Fieldens' appendix to Captain Nares' book, he mentions on August 8th, on the shores of Robeson's Channel, meeting with the young Sanderlings just able to fly but retaining some of the down on their feathers.

is completed. When the young birds do arrive, it seldom happens that any old birds accompany them, or come at the same time. Both old and young, when they first arrive, are easily approached. The young do not moult their quills till autumn in the following year, like all this tribe of birds. On looking over my notes I find the following:—" Feb. 26, saw several Sanderlings on the sands; they were busily engaged feeding; the wind strong, which made the sea break heavily on the shore, and a quantity of white foam was being driven by it along the sands. The Sanderlings were wading amongst the waves which were breaking on the shore, and as it was a bright day, unless you observed they were moving against the wind and in different directions, the whiteness of their plumage made it difficult to distinguish them from the foam. The rapidity of their movements is astonishing: no sooner had a wave broken, than in they ran as far as they could, and picked up such food as the wave brought up; they would then follow the receding wave quite back, till they met the next just going to break, when they retired so as not to be too deep in the water. Probably the whiteness of their plumage is intended for their protection." Audubon mentions the Sanderling as occurring on the coast of America in the same manner as it appears here, but he did not find it breeding. In Capt. Sabine's account of the birds seen at the North Georgian Islands, during Parry's voyage, the Sanderling is mentioned as breeding there abundantly. Surely we cannot be dependant upon America for those which come to us; there must be some breeding places in the north-east of Europe from which our shores are peopled. I have seen the Common Sandpiper swim and dive, and also the Redshank when wounded; and on one occasion, some years ago, when with a friend on a mossy place where Dunlins were breeding, a violent thunder-storm came on; we were close to some of their nests, which we afterwards found. Though the place was pretty dry when the rain commenced, it rained so heavily that the spaces between the small hillocks of heather and moss became running water, across which the Dunlins swam with ease, entering the water without raising their wings. One would suppose birds of the size and shape of those mentioned would **sit too** high upon the water, and be too light in **weight,** and their legs be too long and not offer sufficient resistance **to the** water, to be expert swimmers **and** divers even in times **of** danger.—C. M. **A.,** (Newcastle-upon-Tyne).— P.S. Whilst writing the above I received **a** Grey Plover **in** its summer plumage, much in the **same** state as the Sanderlings are when they first arrive; the **young birds** generally arrive about the **second** week in September. Probably the breeding stations of the **two species** may not be **far apart;** as also those of the Common **Godwit,** Knot, Pigmy Curlew, and Little **Stint.** Little or nothing **seems to be** yet known respecting them.

SPOTTED REDSHANK, EARED GREBES, ETC.

Reprinted from " The Field," May 24*th,* 1862.

On reading *The Field,* **I observe** a notice of two Spotted Redshanks having been killed **lately near** Yarmouth, in the summer plumage. The appearance of this bird in any plumage in England **is** of rare occurrence, particularly so in the plumage assumed by the bird at this season. To persons interested in observing the habits of birds, it would be interesting to know something more about those mentioned—whether they were killed at the sea-side **or at an** inland marsh, and **whether** from a flock; or, if only the **two were** seen, how it happened **the** person who shot them was so fortunate as to procure both, birds of this tribe being generally so difficult to approach. Two pairs of Greenshanks are also **noticed as** having been killed. The appearance of the latter bird **on the** South-Eastern coast **during** its spring migration is much **more common than the** former; but it is naturally a very wary **bird, and it** seems strange that the **two** "pairs" should have been **obtained.** Probably they may have been killed from small flocks; **at any** rate, it would be satisfactory to know how they were obtained. The mention of pairs leads me to suppose the two birds **in each** case were paired; but do these birds ever pair till they **arrive** at their breeding grounds? On the 21st April, 1842, I **got in** Leadenhall-market a fresh-killed Spotted Redshank, about half changed to its summer plumage, and then first noticed the change in the colour of the legs even, of this bird at this season. In winter, and when young, they are similar in colour to the Common Redshanks; but at this season the legs become **dark** brown, and by the time the bird again gets its winter plumage, they again become **red.** I have obtained several in Leadenhall-market in the winter plumage, in the month of November. In this neighbourhood we only meet with the bird very sparingly, and then only in its young or first plumage, in August.* Before Prestwick Carr was drained, it used occasionally to be there at that time, and it is also at that season sometimes on the shores of the Solway. Even when so young, **and** having only arrived from its breeding-station (Lapland or some other northern region), it appears to be then very wary; and a person who has killed two at that season told me he had great difficulty in approaching them; one he killed at dusk by hiding himself and imitating its cry. The Greenshank: I have observed and procured the young of the year as early as the end of July; and although much more common during its autumnal migration (all those which I have seen **at** that season being young birds of the year), sometimes acquires **its** winter plumage, and remains at places suitable to it during the winter—at least, I have obtained them in October and November, and again in February. I cannot remember ever seeing one killed near here having the spotted back, breast, and neck, which the bird gets in summer. I have often seen Greenshanks in Leadenhall-market in the month of April, in summer plumage. I see also a notice of Temminck's Sandpiper having **been killed.** In this neighbourhood the young birds of this species **occasionally occur in** the early part of September, I have noticed it between the 2nd and 11th of the month. They appear to remain a very short time during the autumnal migration, and are never seen in the winter plumage. I have one old bird in its summer plumage, killed by the river side on the 25th of May, 1845; it was alone at the time, and I have seen one other killed at

* On the 20th August, 1878, I was fortunate enough to fall in with **two of these elegant birds at Holy Island, one of** which I shot.

the same season of the year. Two pairs of Eared Grebes are also mentioned as having been killed, it would be interesting to know how they were obtained; the habits of these birds are such as would render the obtaining of both "pairs" a matter of some difficulty, as at this season the weeds on the ponds are getting up and the birds dive so readily. I do not mean at all to doubt that the birds were obtained, but to any one interested in the subject some account of the way by which they were obtained, and whether they were actually pairs, would be interesting. Near Holy Island, on the 18th of March, 1851, I shot an Eared Grebe from a flock of five, with a considerable portion of its summer plumage on; the water was rough and the birds were at a considerable distance; when they rose, which they did readily, they were not distinguishable from immature, or female long-tailed ducks, which species was very numerous there at the time. I remember once seeing a bird of this species alive in Leadenhall market in a basket at this season, in perfect summer plumage, and a strange-looking bird it certainly was, as it sat with its yellowish-reddish feathers stuck out from the side of its face. It had been sent from the Continent. I have little doubt but that we will shortly hear of the appearance of some rare birds attracted by the flooded state of the fens, some of the migratory Sandpipers often appearing on their way to their breeding grounds later than this—for instance the Sanderling, Pigmy Curlew, Knot, Little Stint, Common Godwit, Grey Plover. I have known all these birds killed in their summer plumage, and evidently then on their way to their breeding grounds sometimes even as late as the first week in June. The reason one must suppose is that their breeding-ground has probably been covered with water from melting snow, and after the floods which have to subside, and the land has to dry suffiently for them to build their nests on. I think your Caithness correspondent will be found to be right respecting the number of eggs of the Black Guillemot. The eggs are totally different in shape, and in proportion to the size of the bird, from those of the Guillemot and Razor Bill.—C.M.A. (Newcastle-upon-Tyne.)

SPOTTED REDSHANKS, ETC.

(*Reprinted from "The Field" Newspaper, June 7th*, 1862.)

I am obliged to Mr. Fielding Harmer and J. P. (Great Yarmouth) for the information about Spotted Redshanks, Greenshanks, and Eared Grebes. Perhaps Mr. Harmer would say (**not for my information only**) whether the Spotted Redshank has been observed before this year at this season, or at any other. So **many persons** are now paying attention to **the habits** of birds, **and know the** species **in their various plumages, rare birds are not often overlooked.** Punt-guns also reach many birds which could scarcely otherwise be approached. It did not occur to **me** that the birds might be killed from punts. On the 30th of April, 1860, a fisherman at Cullercoats killed a female Horned Grebe in summer plumage. **It was alone;** the eggs in it were considerably enlarged. Being common in Iceland, where it breeds, it might have been on its road there. The eggs in many female birds killed from flocks during the spring migration are often a good deal **enlarged**, though the birds are far from their breeding grounds. It **is** curious why some species of shore birds which winter with us commonly, those which have wintered, appear **to** leave us as soon **as** mild weather comes, and before getting any of their summer plumage, whilst in the month of May flocks of the same species appear **at particular places** in **various states of** plumage between winter **and summer, many in complete** summer plumage, on their way to **their breeding places. One** would suppose that had these birds spent **the winter further south,** they would have migrated **as** early as those which wintered **with us:** it is difficult to understand the migration of birds. **Some** persons imagine that killing **a** rare migratory **bird** prevents the species becoming more commonly met with: this may be so in some cases, but it is **not** in all. The Pigmy Curlew, for instance, comes during September in flocks entirely composed of young birds of the year; some years more rarely than in others, no old birds accompanying them. They seem not to be able to bear the cold, as by the end **of the** month **or the** first week in October all have gone **again,** before commencing to get their winter plumage. I never knew an old bird obtained in its winter plumage, and very few indeed of them are met with at the same places during their vernal migration. Draining, cultivation, making docks, and inclosing wastes do far more towards expelling such birds than shooting. Wherever there are suitable places and food, there the birds will come; how they find particular pools and places, it is difficult to make out.—C. M. A. (Northumberland).

VERNAL MIGRATION OF THE LITTLE STINT, ETC.

(Reprinted from "The Field" Newspaper, May 10th, 1863.)

In an account of the arrival of summer birds at Kingsbury, the Little Stint is mentioned as appearing for a short time during its vernal migration, together with the Dunlin, and other species. Whilst the latter bird is to be found with us, at such places as suit its habits at the various seasons, during every month in the year, the former is a very rare bird except during the month of September, when some young birds of the year generally pass in their autumnal migration, but even at that season it is only found at a few places, and is by no means common. The only specimen in its summer plumage that I know of as having occurred in the North of England I purchased at the sale of Mr. Heysham's collection. It was killed at Brough Marsh June 1st, 1839, by James Cooper, who then lived at Carlisle. His account to me at the time is as follows:—"I shot the Stint just as I was leaving the sands in the afternoon to go upon the marsh; heard a strange note amongst a few Ring Plovers that alighted upon the grass about two hundred yards in advance; requested a friend, who was with me, to pass round them while I lay down at the edge of a creek. When he got within one hundred yards I again heard the note approaching me; thought at first it was a clear-voiced Sanderling, not thinking at the time of a Stint, which I had often heard in autumn, but never at this season before. The note of both Sanderling and Stint is a short sharp chirp, the Stint much clearer than the Sanderling, often repeated when on the wing by both birds. It was passing at about forty yards with the speed of the wind, which was blowing hard at the time: I succeeded, however, in hitting it. I intended to have sent it to you, but arrived too late for the five o'clock train; was just unpacking to examine it when Mr. Heysham stepped in to inquire what news, and on seeing it begged I would stuff it for him, and, as I lay under obligations to him, did not like to refuse. I shall endeavour, however, to send it over for your examination." It would always be useful and interesting if your correspondents would notice when any migratory bird, which is not commonly met with throughout the country, is captured.—C. M. A., (Newcastle-upon-Tyne).

SPOONBILLS SHOT IN SUFFOLK.

(Reprinted from "The Field" Newspaper, May 10th, 1863.)

WILL J. M. be so good as to give some more information about the birds he shot? Will he say whether there was a flock of them, and the sexes of those obtained, which by this time are most likely preserved, and whether they had the bare places on their breasts which birds that have been sitting on eggs get? I think I have got Spoonbills' eggs in Leadenhall Market (taken on the Continent) much earlier in the season than this. It would be worth while trying to find out what these birds were doing away from their breeding ground at this season. If breeding birds, one would expect them to be with their young, and they would scarcely be flying great distances yet. I have sometimes thought, when birds appear at a distance from their breeding grounds, as in this case, that they have been deprived of their eggs or young, and, not intending to breed again the same season wander about, and appear at different times than during their regular migrations.—C. M. A. (Newcastle-upon-Tyne).

SPOONBILLS SHOT IN SUFFOLK.

Reprinted from "The Field" Newspaper, June 6th, 1863.

I HOPE that J. M. will comply with the request of C. M. A. (Newcastle-upon-Tyne), and give us all the particulars he can respecting the birds he shot, more particularly as to the situation on the coast in which they appeared, the time of day or night when killed, and if previously observed in the same locality. In the mean time, I am happy to communicate some particulars respecting them which will probably be interesting to both your correspondents. Through the kindness of Mr. Sayer, birdstuffer, of Norwich, to whom J. M. had sent them for the purpose of preservation, I had the opportunity of handling both birds in the flesh, and subsequently of determining the sex in both cases by careful dissection. Both birds were in full breeding plumage, with a broad, buffy band across the upper part of the breast, and a slight mixture of the same amongst the crest feathers. In one, however, the buff tints were brighter, and the crest feathers altogether longer and more developed, which led me, from merely external appearances, and the fact of the two being killed together, to take them for male and female. The result of internal inspection proved how necessary it is in all such cases to take nothing for granted. Both birds were females, the one with the finest plumage having a larger cluster of eggs, both as to size and quantity, than the other, the biggest, perhaps, about the proportion of a small hempseed; but in neither case did the ovaries exhibit the slightest signs of the birds having paired for the season, and consequently the feathers on the breast showed none of those nesting symptoms alluded to by C. M. A. I opened the stomachs of both, and, excepting a few small pebbles, found them perfectly empty; but the birds themselves were in high condition and very fat, both externally and internally, from which I conjectured that they had either been killed early in the morning before breakfast, or immediately on their arrival after a long migratory flight, but on this point I hope J. M. will kindly enlighten us. When I first saw them the naked skin of their throats and the spatula shaped extremities of their **bills were** bright yellow, and the irides rich crimson. The **finest of these** two specimens was probably three or four years old if not more, and the other not less two years. I have also seen at Sayer's a third specimen, killed about ten days back on Hickling Broad, Norfolk, also a female, which is no doubt a bird **of last** year, being pure white, with scarcely any perceptible crest. The Spoonbill is by no means a rare visitant on either the Norfolk or Suffolk coasts, one or two examples being procured nearly every **year** on their spring or autumn migrations, but more especially in the months of May and June. In this country, of course, the Breydon muds have peculiar attractions, and the Salthouse district has also produced a large proportion of the specimens obtained; **and** even since the drainage of those noted salt-marshes, **they still resort to** the brackish waters which there run parallel with **the beach beyond the high** sea banks. It is now many **years** since Spoonbills ceased to breed in our marshes, yet from their late appearance at times, and being occasionally found in "pairs,"* there is little doubt that if undisturbed they would in some cases do so even now. The high prices, however, offered for all rare local specimens, and the cheapness of fire-arms as a means of securing them, render it next to impossible for any bird more than usually conspicuous to escape instant destruction on our inhospitable shores. Sir Thos. Browne, writing some 300 years ago informs us that "they formerly built in the heronry at Claxton, near Readham, now at Trimley in Suffolk. They come in March and **are shot by fowlers,** not for the meat but their handsomeness, remarkable in their white colour, copped crown, and spoon or spatula-like bill." There is also an old record, quoted by the **Messrs.** Paget ["Sketch of the Natural History of Yarmouth **and** its Neighbourhood," by C. J. and James Paget, 1834], to the effect that "a flock of these birds migrated into the marshes **near** Yarmouth in April 1774."—HENRY STEVENSON, (Norwich).

* Has it been proved these "pairs" are ever male and female? More probably they are wandering birds having no intention to breed at the places they **are met** with. C. M. A.

ANSWER.

SPOONBILLS.

(Reprinted from "The Field" Newspaper, June 6th, 1863.

In reply to C. M. A., there were only the two Spoonbills I shot, and on dissection eggs were found slightly developed in one and more fully in the other, although we thought from the length of the crest of one that it was a male bird. The birds had been nearly four-and-twenty hours where I shot them, and, though I saw them apparently feeding, no food of any description was found in their stomachs. I have no doubt they were about to cross the water to Holland, their usual breeding place I suppose. J. M.

THE COMMON SANDPIPER.

(Reprinted from "The Field" Newspaper, February 28th, 1863.)

A. Y. Z. says he has shot three this month with deformed legs. He is probably mistaken in the species. This bird seems to be unable to bear cold; it arrives here about the middle of April, coming at once to its breeding grounds. Most of them depart in July; before leaving the country remaining for a short time on the coast. Some birds of the year may be found at particular places as late as the end of September, or even into October. It is not very uncommon to meet with birds, especially those which fly in flocks in the manner described by him. The malformation is frequently attributable to shot. Wigeon, Knots, Dunlins, etc., etc., are sometimes found with one leg only, the stump of the other only remaining, and the wound healed over. Where is the Common Sandpiper met with during the winter months? How interesting it would be could we trace a migratory bird for twelve months, and find out the distance it had flown. I once shot a Richardson's Arctic Gull with only one leg, and the bird was quit fat, the injury not having been recent. It was apparently migrating along the coast with others of its species.—
C. M. A.

SCHINZ'S SANDPIPER AND THE DUNLIN.

(Reprinted from "The Field" Newspaper, Jan. 28th, 1865.)

In the letter from "An Old Bushman," in *The Field* of January 9th, on the subject of forming collections of birds' eggs, he says, he hardly believes these are different species. Will he be so good as to give those of your readers who amuse themselves with Natural History, further information as to his doubts. The former has been so very rarely met with in England, that it is not possible to investigate its changes of plumage and general habits. The few I have seen (all Americans) seemed considerably smaller than the Dunlin, slighter in shape, the beak much shorter in proportion, and, in fact, the bird appeared almost intermediate between it and the Little Stint. Perhaps he will tell us whether the former, in its summer plumage, acquires a black breast, as the latter always does. Schinz's Sandpiper appears to be an American bird; the Dunlin (if the American and European races, or whatever they may be, are the same species) seems to be equally distributed on both continents, the Americans that I have seen being similar in marking to our own, but considerably larger. If Schinz's Sandpiper and the Dunlin are the same species, one would expect to find the former resemble the American Dunlins in size, and not be less than the European birds. Probably the breeding districts of Schinz's Sandpiper will commence further north than where those of the Dunlin terminate: the breeding districts of many of these tribes of birds, although comparatively speaking unknown to us, must extend over immense tracts of country, or whence come the clouds which are found of many of them, from our own shores, even to those of India? The great difficulty in proving the migrations of birds seems to be the impossibility of knowing whether the individuals found at a place at one season are the same as those found at any other place at a different season, particularly as we do not even know that their migrations are confined to north and south; and I cannot see what prospect there is of solving the question satisfactorily. Much less attention has been paid to the migrations

of this class of birds than to those of the ordinary summer inland migrants—most of them inhabiting the sea shores, they are seen by few persons only, and those, perhaps, scarcely knowing, or caring to know, one bird from another; and notwithstanding this, they are much more difficult to acquire any knowledge of than the land birds, in consequence of their irregular appearance, and the wonderful change that has often taken place in their apparel during the time of their absence. In noticing the occurrence of the Eider Duck in Essex, your correspondent mentions the bird as seldom seen in this country: it is abundant on many parts of the north-eastern coast during winter as well as in summer.
C. M. A.

SCHINZ'S SANDPIPER.

(*Reprinted from "The Field" Newspaper, Dec. 26th, 1863.*)

I AM sorry I am unable to comply with the "Old Bushman's" request to send him a skin. A friend of his as well as mine (whom he will probably recognize) some years ago had three young birds of the year brought in a whale-ship from Greenland, which were the first I ever saw. I never had a specimen, doubting its right to a place as a British species, and not caring to possess skins of which I knew nothing. I find Audubon observes that he met with the young of the year only at Labrador in August, and he mentions having met with it further south later in autumn and in winter. He omits all mention of a summer plumage, and does not describe the young. The winter plumage, from birds killed in East Florida, he states to be mixed with yellow on the back, and the scapulars with light red. Some discrepancy appears here: the complete winter plumage of all the nearly allied species is plain grey on the back. Macgillivray merely copies Audubon's description. My reason for asking about the summer plumage was purely for information. I was not aware of its being a European species till the "Old Bushman" mentioned it, and afterwards said it was described by

Swedish naturalists. The uncertainty respecting its localities, breeding and identity I hope he may yet have the pleasure and opportunity to clear up. Though apparently nearly allied to the Pectoral Sandpiper **in many** respects, it differs from it in being considerably smaller, **the** latter bird exceeding the Dunlin **in** size. I have one of these American wanderers, **killed on** the Northumberland coast, June 27th, 1855. At that time it should have been breeding. During the autumnal migration one can suppose it possible that a chance individual of a species breeding **in** high northern latitudes in America might get too far to the east, and thus reach the European coast, and find its way here in the autumn without **directly crossing the** Atlantic; but it seems to me more singular to meet with a straggler on its way back again. I might add, that by attention you may in time, **from** the condition and colour of the feathers of this class of birds, tell within a month or so of the time of year when they were killed, whatever age they may be. After they have once moulted to a complete winter dress, which they do the autumn of the next year after they are hatched, I believe it is not possible to distinguish the age; it is sometimes possible to test the correctness of statements as to their history, date of being killed, etc. C. M. A.

Probably Schinz's Sandpiper really was never found in Europe. The Swedish birds are only Dunlins evidently.—C. M. A.

ON THE ALTERATIONS TIME AND OTHER CIRCUMSTANCES MAY HAVE MADE IN SOME KINDS OF ANIMALS.

Who can tell what some of the different animals apparently now existing in a wild state in various parts of the world are, and what they would have been had not man interfered directly or indirectly with them? Take, for instance, what is called the **Wild Boar.** Who can tell **where his first predecessors** were placed, and what they were like? Is it like cattle, and sheep, **and goats,** one of the animals probably created especially **for** man's use, and were they created semi-domesticated, if not totally so, and placed where they were so that man might take them to be useful to him?

May **not** all the Wild Boars, distributed over the various parts of the world they are **now found** in, have descended from one **common** stock, and, **having been taken by** man to other countries, **centuries** ago, and allowed **to stray, or perhaps having** strayed of their own accord, have become **naturalized, and time** and circumstances altered their appearance, and who **can tell to what** extent?

The same remarks apply to cattle, **sheep and** goats, which I think undoubtedly the Creator in his wisdom placed in Adam's way domesticated **or nearly so.** If it were **not so, how were** they completely **subdued** so quickly from **a** savage state? No wild animals can be so domesticated at the present time; all experiments to reclaim other species, with few exceptions, **and** make them equally **useful to** man seem **to have** failed. Man, having transported these to nearly all parts of the world, during so many centuries, probably **some** of the apparently indigenous species found **wild in** other countries, are only **the** descendants of those **taken** thither by him, and which, **having** escaped, have colonized themselves, but which are now, under different circumstances, as food, climate, and time, so altered, as not to be **recognized as** the **same** species **they** may perhaps have sprung **from.**

It seems certain that comparatively few species **are** becoming extinct at the present time; **those** which have done so for some recent centuries, have been apparently singularly local, and it would appear an All-wise Providence has **considered** there was no further use for them. Generally if there **is** occasion for a species, Nature takes sufficient care to have it in numerous localities, so that, in case of failure in one or more of them, in others the supply is kept **up.** One **might ask,** what wild species can man now make available to his use? The Common Fowl, probably, was put in man's way intentionally semi-domesticated, and afterwards taken by him centuries ago to various parts, and in suitable places for it perhaps it strayed, and for what we know the different varieties found wild in the several localities in which they exist may have all sprung from the same original stock, but which may have been very different in appearance from any existing at the present time, but still be the same species.

How admirably adapted the fowl is to man's use in every respect: look at the eggs alone, and the easily reared young, so totally different from what might have been expected from such a bird—the natural shyness of the tribe to which it belongs entirely gone. Now the Pheasant, notwithstanding the length of time it has been domesticated, never entirely loses its shyness; individuals may, but the species does not, and the young are very much more difficult to rear, clearly showing it was not intended to be turned into an entirely tame bird, and be useful to man in the way of the fowl. Similar remarks apply equally to the Pheasant and the varieties of it found in the different countries, but probably not including any but those closely allied in form and general appearance. Leaving the birds, but on the same subject, as regards Camels, I think it is apparent they were originally created for man's use, and also Horses. What was the original of either? What were the originals of our Dogs and Cats? May not some of the wild animals now found in various parts of the world, and which are considered species, have in like manner descended from escaped animals, and in time have nearly lost all trace of their originality. I think if such animals produce fertile young when crossed with our domesticated animals, this would prove their identification as species.

Another little animal seems to have been created for man's use, the Rabbit. Why was it selected instead of the Hare, which is so much larger? The Hare, useful as it is, often almost semi-domesticated, seems disinclined to submit to man's further interference; the Rabbit, on the other hand, has done so, **as it** was intended it should, and probably in consequence it **was** created with far greater powers of reproduction in order to make it more useful to man. Our children have kept wild Rabbits, quite tame, as pets for years, in wire cages without bottoms, so that when put on the grass they feed and retire as they please, but they have always been kept separate. They like to be carried about, and to run about a room; but if they do get out outside it is no **easy** matter **to get hold of them** again, **perhaps** more from **their** being frightened than **anything else, they move** so quickly **and** there is literally nothing to catch hold of.

Now we come back again to the birds. What a most valuable **fellow the Turkey is:** no doubt it was intended specially **for man's use** domesticated. The Pea Fowl, and Gold and Silver Pheasants, seem to have been created specially for man as ornaments, and as **such only**, as though they **live and** thrive **well** with his **care—they can** scarcely be called **useful.**

We now come **to Geese and** Ducks. Can any one doubt but that the Grey Lag Goose and the Mallard, the origin of our domestic birds, were created for our special use. No other species of either genus has ever been reduced to a state of domestication like them. **The** Goose has **not** changed much from its typical **form, many still retaining the** original shape and even colour. **The Duck has changed perhaps** rather more, but the Drake al**ways** retains its typical curled tail feathers.

There is one thing which has always struck me **as remarkable which is** this, the similarity in the plumage **of the true wild** Mallards; **though no two are exactly alike, the wild** birds always **show** the true plumage **of the** species. One would think that in some places the wild birds might sometimes pair with tame birds, and that a pied plumage would sometimes be the consequence in the young, **and** that they might become **wild and** migrate and travel about as Nature directs and **teaches the wild** birds, and show variety in the apparently wild birds, but they **do not seem to do so.**

I am quite aware of the varieties amongst the half-bred birds which fly about ponds in parks and other places seldom if ever straying far from home, and if by chance they do returning shortly; but the true Mallards always seem to retain their true colour, finer shape, smaller feet, and more chaste feathers. Even if you could mistake them when swimming or flying, the wild birds are unmistakeable when you take them in your hands, **the true wild birds during** a year probably sometimes flying over a quarter of the globe.

The Rock Dove was evidently another bird intended for man's use: **if it were** not specially, why **was not the** Cushat, a much larger, and, I think, handsomer bird, chosen? **but** it cannot be successfully domesticated, which is quite the reverse with the Rock Dove. No bird is more easily kept than it in the various forms produced by domestication, during probably centuries.

The remarks I have made about the Mallard seem the reverse with it: this Pigeon becoming wild seems not quickly to revert to the original colour, as, from the accounts we have of it at its breeding places, varieties seem often to occur; probably these are caused by domesticated birds straying with those which are wild. Nature however in general approves of uniformity, and probably in time, unless other tame birds became **wild in like** manner, the natural colour would in time become constant again, **as** it originally probably was.

Now for the Canary; and perhaps you may smile at its insignificance. If it was not intended to be a companion to man, why **can** he not domesticate any other species of small bird so as to make it a similar friend? He cannot, however, with all his wisdom and care: he may, as I said before, succeed with individuals, but he cannot with a species.

The Grey Lag Goose and the Mallard were the largest, and perhaps the best to eat of their tribes, and they alone have become useful to man domesticated. He no doubt long ago tried various allied species and failed with them. Why should this have been so unless some rule beyond his conception had ordered things to be as they are? It is different with the Pigeons; the one domesticated is neither so large nor so good as the irreclaimable Cushat. Why is this? These are our native birds; but singular it is, that we should be indebted to America for the Turkey, and to the East for the Fowl, Pheasant, and Peacock.

I do not for one instant mean that man cannot make pets of hosts of species, and with care successfully keep them, but that is not the point. Can he make them generally and extensively useful to himself and others?

THE GOLDEN PLOVER.

(Reprinted from "The Field" Newspaper, February 7th, 1863.)

S. S. S. doubts this bird being nocturnal in its habits. Macgillivray, who gives a good account of it, says: "In the Hebrides I have often gone to shoot them by moonlight, when they seemed as active as by day, which was also the case with the Snipes; but I seldom succeeded in my object, it being extremely difficult to estimate distances by night." I have often watched them in the evenings coming to feed on the worms and insects which come out of the ground at that time. The large flocks which frequently remain at rest, if undisturbed, in the middle of the large fallow fields during the day, in the evening separate into small parties, and scatter themselves over the country; but they do not disperse till it is almost too late to see them, and they generally fly near the ground, and with great velocity. Often, when walking in the dusk where they are feeding, they seem to let you get very near them by the cry they utter when they rise, but it is difficult to see them; and, as they often rise singly, it is almost impossible to shoot them. The writer above alluded to says the Plovers in the Hebrides do not migrate. Now, although Plovers may be found there during the whole year, it is more than likely that those which have bred there do migrate, and that those which are there in winter have migrated also. It is difficult to suppose that a species is not migratory at a particular place, when it is migratory at other places, and when great numbers do migrate from Iceland, Sweden, Lapland, and other countries, where they breed and spread themselves **over such** great tracts of country, and get so far south in winter. **The** Golden Plover **breeds on many** of the high mossy grounds in Northumberland. When the **flocks** first come to the lower grounds, which they often do in July and August, they are sometimes not very wary; the flocks are then chiefly composed of young birds, the old birds being then in the moult; but, at that season, they are not in very good condition. As a general rule, birds of this tribe, when breeding, seem to get a plumage making them less conspicuous; that is, more nearly approaching **in colour** the ground on which the nests are. The reverse **seems to be the** case with the Golden and Grey Plovers. Can **anyone give** a reason why they get the beautiful black breasts, **encircled** with a white border? Can the male become so conspicuous in order to divert attention from the female whilst she is engaged with her eggs or young? In watching for Plovers in the evening, when they fly close to the ground, you may sometimes, by stooping, get a shot at them by thus getting them between you and the light. Audubon gives a most extraordinary account of a day's Golden-Plover-shooting, which may perhaps be interesting to some of our readers. It is as follows: "While at New Orleans, **on** the 16th March, 1821, I was invited by some French gunners to accompany **them** to the neighbourhood of Lake St. John, to witness the passage of thousands of these birds, which were coming from the north-east and continuing their course. At the first appearance of the birds early in the morning, the gunners had assembled in parties of from twenty to fifty, at different places where they knew, from experience, the Plovers would pass. There, stationed at nearly equal distances from each other, they were sitting upon the ground. When a flock approached, every individual whistled in imitation **of** the Plover's call-note, on which the birds descended, wheeled, and passing within forty or fifty yards, ran the gauntlet as it were. Every gun went off in succession, and with such effect, that I several times saw a flock of a hundred or more reduced to a miserable remnant of five or six **individuals. The** game was brought **up** after each volley by the dogs, while the masters were charging their pieces anew. This sport was continued all day, and at sunset, when I left one of these lines **of gunners,** they seemed as intent on killing more as they were **when I** arrived. **A man** near the place where I was seated had killed sixty-three dozen. I calculated the number in the field at two hundred; supposing each to have shot twenty dozen, forty-eight thousand Golden Plover would have fallen that day."—C. M. **A.**

WILD GEESE.

*(Reprinted from " The **Field**" Newspaper, April 9th,* 1864).

Mr. Wright observes (March 22nd) that at that time hundreds of "Common Wild Geese" **were** at Strensall Common, near York. In England, we have four species of **Grey** Geese which visit us: the Common Wild Goose, the White-fronted, the Pink-footed, and the allied species with dull orange legs. Of these, the (so-called) Common Wild Goose is at the present day much the most uncommon—so much so, that several seasons pass over without any turning up. Probably in former times, when extensive marshes existed and population was scanty, this species would be the one residing the greater portion of the year here, breeding in suitable places, and the eggs **being** easily obtained, and the bird readily adapting itself to domestication, **from it** has descended the tame geese of our day. The White-fronted seems to be more nearly allied to the Common Wild Goose than the other two, their beaks are larger in proportion, in comparison to the size of **the** head, and are all pale coloured. The White-fronted **Goose (as well as the Common Wild Goose to a much** less extent) **are** exceptions to nature's general rule, in this respect, that when old, each individual varies in its markings on the breast—patches of black feathers, varying in size and shape, and the markings on the two sides of the bird not corresponding, being general. **The** Common Wild Goose has traces of the black feathers only, whilst in the White-fronted every **variation is** found, from **approaching** black, to few **feathers only of** that colour. The **Pink-footed Goose is, I** believe, the **species** noticed by Mr. Wright, **or it may be** the orange-legged **one.** These two species are sometimes met with together. I have heard of their having been killed from **the** same flock. Although they **are** easily distinguished from each other when recently killed, in preserved specimens, when the beaks and legs have faded, it is difficult to distinguish them. In the pink-footed **bird the neck** is shorter, it is generally greyer in colour, and the beaks when fresh have an irregular-shaped spot of pink on a black ground colour, the orange-legged bird having a similar shaped spot of orange. The four species, although spreading themselves over great tracts of country, are nevertheless local in **their appearance,** flocks of certain species apparently stopping at **particular** places during their migrations. All the species when **young and** fat are good eating when not kept too long, but, like the rest of waterfowl, they should be eaten fresh : if kept till the fat begins **to** change colour and turn greenish, which it soon does, they get **a** disagreeable oily taste, many persons erroneously imagining **this** to arise from their feeding upon fish, which it is perhaps **hardly necessary to say** geese do not. **The Common Wild** Goose **is** an early breeder in a domesticated state, and has **young before ths** pink-footed species departs to its **breeding** ground. From its remaining **so** late we may almost conclude that **it** goes far north, and breeds in morasses and swamps not earlier ready to receive it. It sometimes remains till May, and **in the** middle of April in some places it is found in great numbers. **Some short** time since I made some remarks about the Common Wild Goose. Many of your readers are doubtless aware of the different species ; some may not be, and to those these remarks may be instructive.

It is not difficult to distinguish the old from the young birds when you have them together; each feather of the young bird is smaller than in the old, and in the old birds the general colour is purer and the feathers seem more compact than in the young birds.—C. M. A. (Newcastle-on-Tyne).

WIGEON, ETC.

(Reprinted from "The Field" Newspaper, Dec. 6th, 1862).

THE state of plumage in which a male Wigeon is described by "Astur" is remarkable, considering the season. The only way to account for the quills being as described is that, from some accidental and very unusual cause, the bird had not completed his moult at the regular time, and had migrated here before doing so. If he had been caught by a severe frost in fresh water, he would have been in great danger of being made a meal of by a Fox. I think, as a general rule, the drakes of the Common Wild Duck, Wigeon, and other allied species, cast all their quills at the same time; this they do in July, and the birds continue skulking about amongst the reeds and coarse covert until they grow again. At that season all drakes are in their plain plumage, which nearly resembles that of the Ducks, and when they are met with they are often taken for flappers. During the time they are in this apparently helpless condition, nature takes care they shall have plenty of food, without their being obliged to travel to any distance. Their wings grow again before the approach of autumn, and by the time it is necessary for them to migrate—either in search of food or for some particular food they like—they are in full vigour, and able to undertake their autumnal migration. The young of all Ducks, Snipes, and, indeed, of all birds, get their first primaries secondaries tail and other feathers, in most regular order; all the feathers commence to grow simultaneously, and consequently they all arrive at their full length about the same time. In true game birds the young commence to moult their first primaries, secondaries, and tail feathers so soon as ever they have come to their full length, and before the birds get into full plumage (which they do before winter) they change every feather. The **young of** almost all other birds, except the small land birds, **retain their** first primaries, secondaries, and tail feathers till the **autumn the of** the following year, though sometimes some **species,** Snipes, Woodcocks, **Godwits,** and others, change some of **their** tail feathers before winter (some, perhaps all of them); sometimes the **two** middle feathers, sometimes those at the outside only are changed. The first feathers a bird gets are smaller (both shorter **and narrower**) than those it gets when they are cast, and sometimes, in consequence of the two middle feathers, or the two outside feathers only, being renewed, the shape of the tail is altered for the time. **I have often** wondered why the Wigeon is so long in getting **its** full plumage; **the** Mallard, both old and **young,** are in full plumage by October or November, whilst the **young** drake Wigeon seems to get a plain plumage after **its first plumage,** and only changes by slow degrees to **his perfect plumage,** many of them not having obtained the **white patch on the wing** before leaving us in March, April, and **even May the following** year. The old Woodcocks **seem** to **get completely through the autumnal moult** before arriving in **this country;** the plumage of the old and young Woodcock **is very similar, which makes it** difficult to distinguish them, much more **so than the Common and** Great Snipe.—C. **M. A.** (Newcastle-upon-Tyne),—P.S. In illustration **of** what I have said **respecting the difference** in the **size of** the **first feathers of** the young **bird of the year, and** those **next got by it,** I enclose two feathers **from** the **tail of a** young Woodcock shot **last** week, the larger and more pointed one still moulting, having been taken from the outside of the tail on one **side,** the smaller feather having been taken from the outside on **the** opposite **side** of the same bird.

THE EIDER DUCK.

(Reprinted from " The Field" Newspaper, May 27th, 1865.)

A few weeks since an account of this bird appeared in THE FIELD. A few weeks previously, in reference to a remark as to its rarity in England, I mentioned its being common, both in summer and winter, on the Northumberland coast. Your last correspondent seems to say it is not so. Now there are some birds becoming scarcer by reason of their peculiar grounds being altered, while others are getting scarcer from being wantonly destroyed; but the Eider does not seem to me to suffer from either cause; and I do not think your correspondent gives Northumberland due credit for the possession of this fine species. True it is, it is not very frequently shot; but this is not on account of its rarity, but from the difficulty in getting near it. The Eider does not heedlessly fly, in the apparent security of evening, to certain destruction, as the Mallard and many other Ducks do often. Neither does he come into the shallow, smooth water, or sit upon mudbanks where he can be approached with a punt, as the Sheldrake does; but he seldom even flies within shot of the shore, constantly keeping the sea amongst the rocks, often in scattered flocks of a dozen or more, where, in fine weather, he is unapproachable, and even in rough weather it is rough and very uncertain work to get him; and as he feeds on dogcrabs, and his flesh looks very coarse, I do not think he would be much relished as a delicacy. Possibly the reason of its not being found further south is the absence of bays, amongst rocks and rocky islands, where they might find shelter from weather, security from their enemies, and food, during the most violent storms. I hope and believe the year is far distant when it will be a rare bird on our coast, particularly as the islands on which they breed are preserved, and the rocky nature of them renders it unlikely they ever can be made available for the requirements of man. Had these islands been of sufficient extent, there is every reason to suppose the bird would have been as abundant with us as it appears to be on the rocky island-bound coasts of Northern Europe and Northern America.

What has appeared singular to me is, that in October you see the young in flocks; further in winter, and more particularly in spring, most part, if not all, you see are mature birds. Now, what becomes of the young during the time they are undergoing their change **of plumage;** it is not supposed, nor is it likely, **they** acquire their full plumage the year after being hatched; but, as this seems the southern limit of their migration, **what** does become **of them?** The white plumage of the old Drakes after the breeding season becomes spotted with black, and portions of this plumage are visible till about the end of October; but they do not appear to get any of the rust colour similar to the plumage of the Duck. It seems to me possible that the birds recorded as King Ducks killed in England may be merely faded birds of this species, **as it is** evidently a much more northern bird, and **apparently does not migrate** far to the south in **winter.—C. M. A.**

SPHINX CONVOLVULI.

Reprinted from "The Field" Newspaper, Sept. **16th,** 1865.

Sept. 7. One of my children has brought **in his** cap what he called "such a large butterfly." It is **one** of the above, fresh and perfect; he had found it **on** wall **close by.** What an elegantly formed insect it is! and how beautifully the pink and black arrangement **of** colour **on** its body contrasts with its, in other respects, sombre ash hue! Its large eyes **appear in** some strong lights as if **red-hot.** This Sphinx, **as well as the** Death's-head and Galii, **in some seasons** seems **to get dispersed more** generally over the country than the Privet, **the Lime, and** the Eyed Willow; the three **latter species,** though much **more** common in **particular** districts, rarely, **if** ever, appear to **occur** beyond those limits (however fine the season), though the food **on** which **these** caterpillars feed is quite **as** generally distributed. Little **seems** to be generally known about the habits of these insects. Are Convolvuli and the Death's-head night-flyers? and do they take long flights or remain in the vicinity where they are bred? It seems strange that the caterpillar of an indigenous insect should feed on an exotic plant. What did the Death's-head feed on previous to the introduction of the potato? The Humming-bird Sphinx was common this summer in the latter part of June. The last year they were common they appeared in May, when the yellow azaleas were in bloom. I remember writing in THE FIELD soon afterwards, remarking that I had been disappointed; the insect had not remained abundant, as I supposed when the breed was once got up again it would so remain. Caterpillars that year were to be found, and as the species appears again the same year, I had expected, from the quantity in spring, the number in autumn would have greatly increased, but it was not so. The Caterpillars kept in confinement came out in autumn; but at that season I saw but one perfect insect, and the following year they were as scarce as ever. I then inquired how this was to be accounted for. That year those seen in May appeared to have lived over winter, being much paler in colour than those bred in confinement, and paler in colour than those which appeared this year in late June. Is it known at what time the eggs are laid, whether before or after the insect has flown? I presume those that fly early in the season deposit their eggs, which come to perfect insects the same autumn. It seems probable that the eggs laid by the later flying insects may not produce perfect insects till the next spring, and even then only if the temperature is favourable to their coming from the chrysalis. Many of the larger moths, and some of the butterflies, are very regular in appearing at their exact seasons, and only fly at such seasons, and sometimes very short they are, during the year; whilst others, like the Common White Butterfly, appear to come in rapid succession to perfection all the summer and autumn, provided the weather is suitable. I remember a few years since hearing a humming noise in the room which I thought was made by a large beetle, which ceased on the light being extinguished; in the morning, at daylight, I heard the same noise, and on going to see what produced it, I found it was a Humming-bird Sphinx fluttering against the window. The species, perhaps, has derived its name from this noise, as well as from its resemblance to a

Humming-bird in its flattened body and manner of flight. This insect often comes into the house apparently attracted by flowers, or by seeing a light through when the doors and windows are open. I have seen it fly, with its trunks extended, at a small quantity of wet sugar put before it when in the house.—C. M. A. (Newcastle-on-Tyne).

THE SHEARWATER.

(Reprinted from " The Field" Newspaper, June 4th, 1870.)

When on the coast I bought a Shearwater Petrel, which I was told had been drowned by having got entangled in the fisherman's nets. It was saturated with salt water, and dirty. I washed it in fresh soft water, and next morning it was as clean and dry as it probably ever was before. It is the first I have seen recently killed; and what a singular bird it is, apparently approximating to various genera. In form and colour, when lying dead, it resembles a foolish Guillemot in its winter dress; but it is smaller. The plumage, particularly the primaries and wing feathers, instead of being harsh to the touch as in the Auk's, are soft, more like some land bird's, and resemble in some degree those of an owl; the beak, which is black, resembles that of a Cormorant (Willughby remarks this); the feet and legs resemble those of the Red-throated Diver, and the leg bone which joins the thigh bone is elongated as in the Diver. The colour of the legs is singular, the outside being pink, irregulary spotted black, with a hard outline. The wings, which are placed further back than most species, when extended resemble those of an albatross. From the formation of this bird one is led to suppose that it can dive for food as well as take it from the surface of the sea when flying, as other Petrels are said to do; and this bird's having been taken as it was rather appears to confirm this supposition. Its congeners the Fulmar and Stormy Petrels have not the legs placed so far behind, nor so flat; and they in general appearance more nearly resemble Gulls in shape, and from their form would appear to be unable to dive. The question arises as to what this bird was doing on our coast at this season. Hewitson says it breeds in June and July in Shetland; probably it is a regular migrant past our shores from its winter home to its breeding stations in numbers, but keeps far out at sea or flies quickly past, as it is so rarely procured. Earlier dates are given for its arrival at its breeding grounds; but can they be relied on? Some confusion evidently exists as to this bird. It is said to have been common during the breeding season near the Isle of Man, and to have bred in rabbit holes. Now, the Puffin does breed in holes in the ground; but those who have taken the eggs of this bird in recent years describe them as been found in almost inaccessible cliffs, amongst stones. Lowe, in his "Fauna Orcadensis," says "the islanders have to risk their lives for the young," and gives a more lengthened description of this species than he does of many of the birds he enumerates; but at the same time his remarks respecting the risks and difficulties in obtaining birds and eggs, though under the head of this species, must relate to the subject generally. How is it it is so difficult to trace the migrations of so many species of birds which pass our shores to and from their breeding places to their winter quarters? and where are the winter quarters of so many of those that do periodically pass? Willughby describes the Shearwater, and also " the Puffin of the Isle of Man, which he takes to be the Puffinus Anglorum," which he saw taken out of a nest, as "equal in bigness to a tame pigeon." The remainder of his description is taken from birds in the repository of the Royal Society and in Tradescant's cabinet, and which certainly belongs to the Shearwater. The eggs of the Puffin and the Shearwater and the size of the birds being so much alike, I have often supposed confusion has arisen respecting them.—C. M. A. (Newcastle-upon-Tyne, May 20). [The Shearwater, or Greater Shearwater, has repeatedly been taken on hooks baited for fish in Mount's Bay, Cornwall, where, the late Mr. Jackson informed us, it sometimes appeared by thousands. The history of this sea bird is still incomplete, and we are glad to receive any information respecting it. Yarrell's third volume may be studied with much advantage on the subject.—ED.]

THE FULMAR PETREL.

A correspondent, C. M. A. (Newcastle-upon-Tyne), in THE FIELD of June 4, wished for information as to the habits of the Fulmar Petrel, and expressed some doubt as to his diving. I have very constant opportunity of observing them. They dive freely, but in a peculiar manner, following the shoals of sprats or young herrings in great flocks on the wing, and, dropping over them like a hailstorm, dive about 6ft. or 8ft. beneath the surface, using wings and feet, as most sea birds do, but they appear never to dive deep or remain long under water. They are particularly bold and fearless when in pursuit of their prey, and by throwing out bits of fish on a clear day will be kept round a boat, giving capital opportunity of observing their habits.—AQUARIUS.

WATER-RAIL'S NEST IN NORTHUMBERLAND.
(Reprinted from " The Field" Newspaper, July, 1867.)

On July 12 one of my sons was looking for flappers at the side of a lough of about fifty acres in extent, one-third of which is filled with reeds and other water plants, when a pointer he had with him pointed a nest containing seven eggs. He broke one, and finding it recently laid, he put another in his cap and brought it to his brother and myself, who were at the other side of the lough, saying to the former that he had brought him a Waterhen's egg. I knew the egg was a Water-rails, and as I had never seen a nest, we proposed to go in search of it again. This appeared at first a hopeless task, but by retracing the track previously made amongst the weeds it was again found. The nest, which was firmly attached to the stems of the water plants, was about thirty yards from the edge, in water knee-deep, whilst resting on the matted roots of water plants, which bore our weight. The bottom of the nest was in the water, and the nest was raised about four inches above the water, and made of the flattened stalks of water plants. It was quite hidden from view by the arched leaves of carices which waved over it. The seven eggs were apparently the number the bird intended to lay, as they had been sat upon for a short time, and the yolks were very dark in colour, resembling those of the eggs of the Terns. I am not aware of any authentic instance being recorded of the Water-rail having built in this county, though the nest is noticed as having been found in the fen countries not very uncommonly. The Water-rail is one of several species of birds whose habits appear contradictory to the usual migratory movement; with us it is more generally seen in winter than at any other season, and still it is apparently more plentiful in summer farther south. Its movements, like those of other birds, are mere conjecture. Two theories occur to me. One is that the bird may breed in many places, and that many more do breed with us than are ever discovered, in consequence of the great difficulty in finding their nests, or even seeing the birds during summer, when water plants are so rank ; and if the nest is found, the probability of its being taken for that of a Water-hen or Land-rail, as so nearly was the case in this instance. The other, which I think the more probable, is that it really is scarce, but that there may be some scattered places where nests might be found each year. I saw the bird at the same place at the same season last year.—C. M. A.

PRESTWICK CAR.

The 19th of September, 1855, I may say, was almost the last of this celebrated shooting and fishing place. On that day the then Lord of the Manor, Mr. Mitford of Mitford, and a party, came over to see the fish taken from the pool, which was left when the water was run off, never to be collected again. I went with the late Mr. William Brandling, the late Mr. Barrett, then living at Street House, and Captain Shum, then living at Prestwick Lodge. A great quantity of Pike, Perch, and Roach were netted, and we spent a very pleasant day helping to drag the nets, but regretting to see the last of the place where we had spent so many pleasant days.

The Car being a Common, the villagers round about kept their Geese on it, and they were constantly looking after them, and many of the freeholders had quantities of stock also. It was thus much disturbed and was very much liable to trespass in other ways. Tramps came to cut heather for brooms, they damaged the fences and young trees for shanks, numbers of idle people went on Sundays and often set fire to the whins, and at last it was considered desirable to have this prolific naturalist's hunting ground done away with. The Lord of the Manor claimed the shooting: he, living at some distance, gave the deputation to Mr. Bell of Woolsington, a great friend of Mr. William Brandling's, and in consequence the latter had leave to do pretty much as he liked; he being very fond of shooting, frequently used to take me with him.

Mr. Lambert at that time had the adjoining shooting to the north, and he also very frequently kindly took me with him when he went to shoot. The river Pont, which when it overflowed formed the Car, ran for several miles through his shooting grounds, and a most prolific place for Ducks it sometimes was. The first night of the Car being frozen over and a slight covering of snow made many Ducks take the river, which was in many places overhung with bushes, and a walk in early morning was frequently worth taking, five or six couple being sometimes obtained on such an occasion. Some miles lower down the river it also flooded some land and formed another lake which was surrounded by trees and bushes, and was strictly preserved. This formed a most valuable resting place for the fowl when driven off or disturbed much at the Car. The late owner of this sometimes also most kindly asked me to shoot at it, and the show of wild fowl there was often surprising. This retreat would have been a splendid place for a decoy; the late owner long since told me he had had applications to have it turned into one, but he preferred having his Ducks there at that time as they were. Since the Car was drained the great body of the wild fowl have ceased to come into the district. I had numerous opportunities of observing the many sorts of birds frequenting the Car, and my collection of wading birds contains many specimens shot there. It would be difficult to find several of them now in this county at any season. Many times have we watched the flights of Geese and the Ducks coming to, and going from, the Car to feed on the flooded lands outside in the mornings and evenings.

The late Mr. Richard Reay, who farmed Berwick Hill Farm during the time Mr. Lambert had the shooting, perhaps killed more wild fowl than any man hereabouts. The farm house stood on a hill, and from the front door, with a glass could be seen any flocks sitting where a shot could be obtained from the north side, the Car being bounded by a hedge, every inch of which he knew well, and when the place was flooded the fowl were constantly driven up to the edge and within shot of this hedge; and besides this, from the house could be seen the flocks of Geese whilst yet a mile off, coming in for the night to the Car, they having spent the day on the extensive fields further northward. A lane with high hedges on each side ran east and west, and as he almost always kept a large gun in readiness in his house, he had time to calculate the distance to go in the lane so as to be under the Geese when crossing it, and in windy weather they frequently topped the hill sometimes within reach of his house; but to be so fortunately placed as this it requires the shooter to live on the spot, as such good luck is not to be had at any time.

It may be easily supposed the reclamation of such a place as this caused a tremendous alteration in the natural history of the district; no similar place now exists in the North of England; it was in fact a small fen. At times it collected Swans, Geese, Ducks, and Waders unlimited as to species, as well as quantity. The following is a list of such birds as I have known killed there.

January 28, 1848, I find this note.—A very stormy day. At Prestwick Car with Mr. L. (who had the adjoining shooting) and another. The cry of the Wild Swan was heard. Two came and settled on a piece of open water, the only piece there was: we crept to some whins on an island, the ice carrying us, and took positions amongst them. I was eastmost, Mr. L. having gone to where he thought they would fly over on being put up, and the other shooter was to put them up by firing at them. They flew as expected, and on Mr. L. firing at them they turned and came directly over my head; I shot at them, and after they had passed me, one of them gradually became all over blood on the wing, and we saw it was severely wounded. It was snowing hard and we soon lost sight of them, but followed in the direction they had taken, and after going some way on the ice we came to drops of blood, which became thicker as we went on, and at last we found where the tips of the wing had marked the snow. We then found where he had settled by the footmarks, and tracked him for at least a mile on the ice till he had gone to the edge; we then lost his track in the snow in the rough grass and heather, and after looking for a length of time we gave him up, thinking some one must have seen him fall and had got him. Afterwards on being joined by Mr. L. we had another look, and at length found the bird nearly dead, but he had walked a long way from the edge of the ice, and was not easy to see in the snow. This bird is now in my collection. I once also saw a Wild Swan shot with a ball from a flock of nearly a hundred; they were frequently there, but not every year by any means. I saw large flocks on February 10th, 1836, and January 14th, 1847, and I saw some as late as April in 1855.

GEESE.—Pink-footed, the commonest, usually most numerous during spring migration, till nearly May-day. Bean, sometimes shot from the same flock as the former species: Grey Lag generally appeared singly, and sometimes associated with its domestic relatives. I have one killed 28th April, 1855, a young bird of the preceding year. Brent Geese, which so rarely leave the sea; I only know one instance of a large flock which came during a flood and settled on an island. Mr. Reay killed four at a shot, and the rest went straight away. I believe the White-fronted Goose was also occasionally there, but it never came under my notice.

Previous to the drainage of the Car Wild Geese were frequently seen flying about during the winter and spring, even at a distance from it; now they seem to have deserted the locality, and it is a rare occurrence to see a flock during the whole year, showing that the resting place having been done away with they have altered their course. I find two memoranda of Geese being shot when I was present. On the 27th January, 1852, Mr. L. shot one Pink-footed Goose; and on the 20th October, 1854, the bag contained four Pink-footed Geese, shot in the evening amongst other spoils got during the day. Others were shot that I have omitted to take notes of at the time.

DUCKS.—Mallards often abundant. Many bred there and in the neighbourhood, but being a good deal disturbed, after the young could fly well they left and no number were observable till the winter arrival about the beginning of November, especially if the weather was cold, and snow had fallen on the hills, which always brought them, and by this time the water generally got to such an extent it was not easy to drive them away, and they remained till the frost came; but so soon as the frost became very severe not a Duck would be left. I find a note I have January 4th, 22° of frost, travelled the Pont for miles, not a Duck left in the country, edges of the river all ice, no feeding place left; and another note, January 9th, 1849, when with Mr. Lambert; we got eight Mallards at a rise, we saw a number in a hole in the ice which we crept up to. A similar pool left unfrozen and at the same place the Swans came to, and which were approached in the same manner.

WIGEON.—Quantities during open weather, remained from September till frozen out, and again in spring, taking their final departure by May-day, except a single Duck, which I shot on August 2nd, 1852, while looking for flappers, and which had probably been unable to leave at the proper time in spring from some accidental cause. Some of the old drakes are in full feather again by November. I have frequently seen Wigeon there till very late in spring. On the 16th April, 1856, I shot two drakes from a flock of forty or more in full plumage except the wing coverts, which were not white but dusky, those of the young bird. Probably such birds might not breed that season. I have frequently seen them as late in other years. In spring before leaving us the head of the drake gets much lighter in colour, particularly the front. Though young drake Wigeon are so late in leaving as the end of April, and not having acquired the white wing coverts, I think they must breed that year, or why should they all leave us if the young birds do not; there must be a **great number of them** during **summer** somewhere. **It seems difficult to offer a conjecture on the subject.**

TEAL.—Used to come in large flocks in March when the drakes were very quarrelsome, chasing each other and splashing the water about. By April those remaining had paired, but most of them had passed on their migration. In May the drakes were in flocks of five or six together, the ducks probably sitting. There were frequently also Teal in flocks of young birds in September, especially in showery weather, but rarely in mid-winter, if the weather was at all severe. These are invariably young birds of the year. I never met with an old drake with its black beak in **autumn, and** therefore presume they have migrated southwards **early.**

PINTAILS in immature plumage used often to cast up in autumn. I have met with them September 17th, October 20th and 29th, **and** at various other times unrecorded during autumn and early winter only, **often** with Wigeon. I never saw a Pintail between **the middle of** April and the middle of September: they used to come on their spring migration occasionally, and also to Gosforth **Lake: they seem** to rejoice in water they can reach the bottom of by turning **up and** putting their heads on to the ground, their long necks **being** apparently adapted for this special purpose. I remember once winging a drake at the Car in the beginning of April, but losing him by not being able to get through a hedge quickly enough and he beat me, getting into deep water. About a month afterwards I found a clean picked dead bird, but with the feathers partly left on him, and which had probably belonged to the bird I lost.

SHOVELLERS came in March to breed, and generally those which got away left early, as they were rarely seen during early autumn or in winter. I have a drake and duck, both shot by Mr. Barret, the latter on July 26th, 1852, the former was amongst the spoils of a day's shooting, October 2nd, the same year, during a very high flood; it rose from a wet place amongst the heather, and I well remember picking it up, for I have good reason to remember the day, as on it I got a thorn into my shin in getting over a dead hedge, which laid me up for several months. Numbers of Geese and Ducks were then there. This is an old bird regaining its fine plumage from the Duck-like plumage of summer, the wing feathers renewed except the tertials, the beak black. I have seen no other like it in the neighbourhood. I find a note of Mr. Lambert having killed a young drake one evening in December, but I have not the exact year, which was then only commencing to get its drake plumage, having a few of the dark coloured feathers on the head and a few freckled feathers in other parts and with the blue wing. The Gadwall I have before alluded to, as also the Gargeny.

All these, what I call surface feeding Ducks, rejoice in water in which they can reach the bottom with their beaks by turning up when swimming—once do away with that style of country and they forsake it. The same may be said of waders, whose delight is to walk in the shallow water and pick off the insects and other things on which they feed from the plants growing in it, their long legs are for the express purpose,

I remember Mr. Lambert shooting a Smew out of the Pont in its immature dress, and I also remember once having a good view of a small flock of Goosanders, amongst which were mature birds, resting on the edge of the ice after a long frost, which had a pretty effect as the sun shone on them. As a rule, this latter bird keeps to the rapid fresh-water streams and seems to feed much on trout—its near ally the Red-breasted Merganser when with us as invariably seems to keep the sea and its estuaries, very rarely, so far as I have noticed, having ever been found in fresh-water. It is often common about Holy Island from autumn to spring, but keeps much at sea and is not easy to get near.

The drakes of the Wigeon, Pintail, Gadwall, Teal, like the Mallard, all acquire the plumage resembling the ducks soon after the ducks begin to sit, **and all are** totally incapable of **flying** during part of **the** time they are moulting, as they **cast all the** large wing feathers at once, primaries, secondaries, **and tertials,** as do Swans **and Geese,** and the new feathers take about **a month** to grow to their full length.

POCHARDS **and** TUFTED DUCKS were common during winter in open weather in flocks, but they **were** rarely shot, keeping and feeding in **the deep** water and rarely coming to the edge, as they obtain their **food** principally by diving, and Golden Eyes were also common (but not in flocks) during winter. In frosty weather they generally took the river and were often shot, principally young birds: the mature birds used to come in **April on** their spring migration and rest a few days. Herons were **often** there, and I got a Little Bittern shot at the Pont **side on the 31st** May, 1866, within a short distance from where the Car was.

COOTS and WATER **HENS were there** in abundance, and the Little Grebe often in winter, but I never heard of one in summer. Gulls were also often there **of** various kinds, and one year I found several nests of the Black-headed Gull, but it did not breed there regularly, probably they were too much disturbed.

On two occasions I saw Black Terns. June 12th, 1850, when **fishing, two came quite near us; and** again 2nd September, 1852, **when fishing we saw several young** birds, which came within a few yards of us many times, seeming not in the least afraid. I have known the following waders procured at the Car:—**Curlews** occasionally bred there, and were often there in spring and autumn. The Black-tailed Godwit has been got there, but I never observed it myself; Redshanks often bred there, but only came there to breed within the last few years before it was drained; Greenshanks not uncommon during autumnal migration in August and

H

September, all young birds of the year; Spotted Redshank rarely, and in August only, and these young birds during autumnal migration; Ruffs occasionally bred there, mature birds with ruff on, only seldom seen, the last flock in the year 1854, when several were procured during May. Sometimes an old bird before getting its ruff came in April and associated with Dunlins; and in autumn young birds were in some years not uncommon, probably migrating from Sweden: sometimes none came for several years in succession. Knot. I saw one killed on the 20th September, 1839, whilst in first plumage; Pigmy Curlew when it appeared always young in September; Little Stint the same. The first Little Stint I ever saw was shot by my brother so long ago as September 23rd, 1837. Two came with the wind past us on flooded land; one fell winged, and when I waded into rather deep water for it, it was running on the floating weeds. Temminck's Sandpiper in June in summer plumage, and in September for the first few days in young plumage, but very rarely obtained. Dunlins bred, but some years much more commonly than in others, often seen in August and September young of the year, and rarely a flock came also during winter in mild weather during floods. The Ring Dotterel came sometimes, young birds in autumn. The Grey Plover likewise at the same time, I have a young bird of the year killed in October. I never knew the Common Dotterel there, nor the Whimbrel; but as I have frequently heard the latter during autumnal migration flying over as far from the sea (only about six miles), it may sometimes have been there, but still if it had risen it would have been heard; and I only once heard of a Woodcock being obtained—it was marked down and followed. I never knew the Common Godwit killed there. The Green Sandpiper as an autumnal visitor in August and early in September, and I once saw one in April. The Wood Sandpiper, it would appear, occasionally bred there, as mentioned by Mr. Hancock, and the young birds were sometimes shot during August probably migrating from Sweden.* The Common Sand-

* If these young birds had been heard in England, they would probably have also been met with in July, but they appear to come each year, within a few days of the exact time only.

piper was also often seen during summer, it breeding at the burn sides in the immediate vicinity.

The times when most wild fowl came were, of course, during floods in open weather; but in consequence of the shallowness of the water it soon froze over, when all the birds disappeared, but on a thaw coming, and so soon as there was water at the edges of the ice, the Mallards and Wigeon returned.

During floods, also in autumn and winter, in open weather, there were often acres of Pewits and Golden Plovers; but, without apparent reason, sometimes under similar circumstances there were none. In February, 1852, I have a note, that probably in consequence of the very open winter, some thousands of these birds made the small islands left uncovered by water look quite black. Probably they assemble to feed on the worms and other insects drowned out by the floods.

Although Golden Plovers never bred at the Car, I have shot them there in full summer plumage as late as the 21st of April, at which time some would have eggs on the moors: these late flocks would be probably migrating further north to breed, and I find also I have seen them considerably advanced towards their summer plumage as early as the 21st February.

Early in July, even on the first of the month, I have seen vast numbers of Golden Plovers, both old and young, which came most likely from the hills further inland, where they bred, and sit in the low fallow fields on the Berwick Hill Estate, and which was only an easy flight from their breeding grounds.

I have seen hundreds sitting with their heads to the wind. During a great portion of the day, they often sit motionless resting on the ground, except when one or two raise one wing or both straight up as if stretching themselves. You might sometimes with two or more guns divert their attention and get them between them, so that on their rising one would get a shot and thus get a few. At this season the old birds are moulting, and soon after this the young are moulting also, the old birds undergoing their regular complete autumnal moult, when they loose their black breasts and also change all their quills and tail feathers, the young losing their first feathers on the back and breast and getting them replaced by somewhat similar coloured feathers, but rather larger than those cast, and less distinctly marked.

In consequence of the rapid moult and the regularity of it amongst the individuals, and the number of them, quantities of feathers lie about wherever a large flock has remained any time, the primaries and tail feathers being those of the old birds and much worn, they having served the birds a whole year, and in the birds of the preceding year from the time they first got them.

In calm weather they move little during the day unless disturbed, but during windy weather these large flocks fly about much more, and often whilst you were wading knee deep in the great extent of flooded land, these large flocks would come very scattered right at you at a great pace and give you a pretty shot, though you could only get a single bird if you could hit it. I have often seen them fly over the dry parts of the Car also in the same way where it was rough heathery peaty land, and often though you saw the bird fall to the shot you could not find it, particularly if carried by the wind, so difficult were they to see in the rough heather. By the end of September the old birds had lost all trace of the black breast and their wings are again nearly complete, and soon after this the flocks are not nearly so numerous, many of them having got their perfect wings and probably migrated southwards.

Snipes if left alone bred abundantly; they only were at their breeding ground from April till September, after which the old birds having got through the moult, they left following the young, which probably went south soon after being able to fly well. Prestwick Car was the only place I ever saw what you might call Snipes at home, and there they certainly were in spring, summer, and early autumn. In winter scarcely one was to be seen, and then not on the Car itself, but on the runner sides. The regular migration of breeding Snipes came in March, and they bred there in numbers, or would have done if they had not been so much disturbed. The place was bad enough in those days; had it not been drained it would have been twenty times worse now, what with railways and other enlightening processes which accompany the progress of civilization. What has this to do with Snipes? it might be asked. But it has a great deal, as in many places it has improved them from the face of the earth entirely where they used to be in plenty. All I can say is they have afforded us many a pleasant day, and sometimes in autumn we could get from fifteen to twenty couple, besides a few Teal, Plovers, and sundries. The best time for them was early in autumn, when not only those bred there were to be found, but others from the moors, when food begun to get scarce, as it then does, came down to the flooded land. At such times acres and acres got covered with shallow water, through which grass came and rested on its surface. On this grass Snipes, Plovers, young Ruffs, and various other birds walked about and fed. The Snipes were awake and moving sometimes all day long, often four or five together, and sometimes mixed with the Plovers and other birds.

Snipes lay in April, but it would seem they will soon lay again if the first eggs are taken or destroyed, as one often meets with young birds not able to fly in August, whilst the young of the regular time should be able to fly in the end of June.* Young Snipes are easily recognisable from the old in August; they are symmetrically marked, the edges of the back feathers being narrow and bright. They retain this plumage till the autumn, when they begin to get new feathers on the back edged with broader marks, being the plumage of the mature bird; and gradually, as winter advances, they get their winter plumage, resembling the old bird, but they do not cast the wing feathers till the following autumn. In August the old birds moult their quill feathers, and they are often so much in the moult as to be unwilling and almost unable to rise; these are often taken for young birds but the difference is easily known; the wing feathers of the young bird

---

* I find notes of having found a nest with four eggs on the 19th April, and also a nest at Sweethope Lake, on the 10th August, 1840, containing two eggs—these probably would never be hatched.

grow all at once and are of corresponding size; the old bird moults gradually, and the wing feathers are of unequal lengths. I think it very likely the old birds migrate so soon as they are able to fly well, and not only they, but by the end of September I think it probable all the Snipes breeding and bred in this country leave, as it often happens none are to be found for some time in autumn, but after a time some come and these probably come from a distance with the Woodcocks and Jacksnipes. Many of the young birds so soon as ever they are able to fly well leave the places entirely where they have been bred long before the old birds have got through their moult; young birds may be known during winter as they retain the wing coverts of the first plumage till the autumn of the next year.

During summer Snipes used to be flying about all day making their singular drumming noise with their wings, and also bleating away, reminding one of their name,—heather bleater. It is surprising what a common bird the Snipe is, notwithstanding the improved state of the country and the immense numbers killed each year. In suitable places it is still numerous, and probably though driven from many of its former localities, it will continue to be abundant. It would be a curious problem to find out whence Ireland is populated with this bird—are they indigenous or do they migrate from more northern swamps? We appear to be exhausting the sea of its oysters; will it be possible to exhaust the swamps of their Snipes, and which would be of the most importance or most regretted?

Within the last thirty years Snipes used to be commonly found in dirty oat stubbles where water lodged and remained in the furrows; nowadays if the rainfall is ever so great the land is so completely drained, the water never remains long enough to attract them; at any rate they are seldom now found in such places. While watching Wild Ducks at night when none have been flying, I have sometimes shot Snipes at moonlight by the water's edge when they alighted between me and the moon.

Black Game were sometimes at Prestwick Car, but it was too damp for game of other kinds.

When the Car was partly flooded in autumn, some of my companions used sometimes to ask a stranger to have a day's Snipe shooting. Drains had been cut in many places to dry the pasture in summer for the eatage, and a sort of grass used to grow which floated flat on the surface of the water; where this grew the bottom was firm, where the drains were there was no grass, and consequently anyone knowing the character of the place avoided the water: the drains, even when there was no flood, were full of black mud; the joke was to go in such a way as to lead the stranger into a drain, and in consequence giving him a bath, and then having a good laugh.

Sometimes when either fishing or shooting some of the party would be taken in the boat, when there was one, to an island on which grew some willows and dense herbage, a likely place for Teal; the party having disappeared to go round, on his arrival at where he started would find the boat gone,—sometimes it required not a little difficulty to wade across again without getting much deeper than one cared about if carrying a gun and ammunition or a fishing rod; sometimes a little temper was shown, but it all ended in jokes. These were pleasant days the remembrance of which must soon pass away, altogether, alas, how few of the participators are left—most of them have disappeared more completely than the Car itself.

Besides the birds, Pike, Perch, Roach, and Eels, and hosts of shells, plants, and insects were located there. The birds having wings escaped annihilation and have gone elsewhere; the fish and shells of course were destroyed entirely, and many of the insects and plants also, in consequence of the entire change, but without doubt another set of animals and plants has taken the place of their previous occupants, and certainly are of more use to man under the altered conditions, but hardly so interesting to the naturalist.

LITTLE BITTERN, ETC.

(*Reprinted from " The Field" Newspaper, September 1st,* 1866.)

Mr. Surtees, of Benridge, has given me a male Little Bittern, in perfect plumage, which was shot by his keeper on the 31st of May last: it has been preserved by Duncan, of Newcastle-upon-Tyne. Bewick's figure of this bird was taken from one procured within a few miles of the same place, and on the same stream, upwards of half a century since, a little earlier in the same month. This reminds one that when a bird has been once met with, another of the same sort may be found not far off at some future time. This species, though a regular summer visitant to some parts of the Continent, appears always to have been a mere periodical straggler in England, and not one of those made more scarce by being destroyed or driven away in consequence of the alteration of the face of the country. It is not noticed by Willughby in his work, and had it been of frequent occurrence in his time it would hardly have escaped his observation, notwithstanding its retired habits. Some persons have suggested that the Little Bittern may sometimes breed in England; but as attention has been paid to Natural History for so many years, and by so many persons without the nest or unfledged young having cast up—the only conclusive proofs—the probability is that it does not. Several migratory species of birds take very long flights immediately before their breeding season commences, and also immediately after it has finished, the consequence of which is that we find some species on our coasts sometimes as late as the first week in June, which have never been found breeding in this country, and must be then far from their breeding grounds; and we find others again returned to England by the middle of July. The Whimbrel, a strictly migratory bird, neither spending winter nor summer with us, though many of them do not go so far north to breed as some other species, passes England in the spring till June, going to breed, and some repass from their breeding grounds as early as July. The only way apparently to account for this is, that those which pass earliest do not go so far north, but breed earlier and return earlier than those which pass later in the spring, and which in consequeuce return later in autumn. Different sorts of birds vary in the length of time required for the purpose of hatching and rearing their young—that is, before the young are capable of flying well and taking care of themselves. Wading birds—as the Curlew, **Golden Plover**, Snipe, Common Sandpiper, Dunlin, and Redshank—require, I think, about eight weeks from the time the eggs are laid till the young can fly well, and immediately after this they are capable of flying any distance. At this season the young of these species can fly better than the parent birds, which have commenced to moult their quills. It seems clear that these birds only produce one lot of eggs in the season, unless the first are destroyed.—C. M. A. (Newcastle-upon-Tyne.)

THE LITTLE BITTERN.

(Reprinted from " The Field" Newspaper, October 13th, 1866.)

In THE FIELD of August 25th Dr. Bree notices the occurrence of a Little Bittern, killed near Colchester, on the 13th of the month. Prior to its publication I also had forwarded to the editor a notice of a Little Bittern killed in Northumberland on the 31st of May, with some observations on the habits of the bird, and they appeared in THE FIELD of September 1st. On September 8th Mr. Marshall suggests, from Dr. Bree's notice, that the bird breeds in England. Dr. Bree says in his bird, a female, there were two small eggs for next season in the ovarium. Is that what Mr. Marshall draws his inference from, and is there not some mistake? I have sometimes examined birds, and in females, after the breeding season, have found a considerable cluster of minute eggs all the same size, and often so small as to render the determination of the sex extremely difficult, if not impossible. At any rate, if this bird would have bred in England had it not been killed, it would have had to spend the winter in some distant country, probably Africa, as it could not outlive a long frost here. Now, thirty years ago it is admitted the species appeared no more frequently than it does at the present time, when there were a much greater quantity of suitable places for it. Many naturalists are anxious to be supposed to have found out, or to suggest, something new; hence surmises. Dr. Bree quotes from Yarrell two Little Bitterns, one a young bird, which was shot on the banks of the Thames in 1828; the other, which was only seen, it is stated was in the same plumage. Now, will any man undertake to say what state of plumage an uncaptured Little Bittern is in when seen on the wing, or even on the ground, at twenty yard off or less? It is of no importance to me raising the question as to the birds breeding in England; but, as I before remarked, the only convincing proofs are the nest and eggs, or unfledged young. I cannot believe any wild bird would make a nest unless paired, or lay infertile eggs; and my theory is that the bird is, and always was, a straggler in this country, lost in its migration. Montagu's notice of a female in May, 1808, which he says contained numerous eggs in its ovary, some of which were large enough to induce a belief that it might have bred if it had not been shot, is more feasible; still, in spring all female birds that are going to breed appear to have in their ovaries, before they are paired, a cluster of eggs in a forward state, many more than they will lay that season (except in the most exceptional cases, such as being deprived several times of their eggs), and this occurs in birds which have long journeys to take before arriving at their breeding places. I believe the eggs come to maturity very soon indeed after the birds pair. Does anyone suppose that if Little Bitterns were never killed in England they would become commoner than they are? Such a supposition appears to me simply absurd. But even suppose they would become common, if they were never seen what would it signify? I do not want them killed, and am as sorry as anyone can be to have them or any other bird wantonly destroyed. It is probably owing to the number of eggs coming forward in spring that the Pewit is enabled, on being deprived of its eggs, to lay again so speedily and to keep laying for so long a time. Its having this power calls to my mind the difference in allied species: why is this not so with other species to the same extent? It is often supposed that, could we have all the species that ever were together, a natural systematic arrangement would be apparent. My observations lead me to the conclusion that the more we see the more confusing and perplexing the whole matter becomes. It appears evidence is not present that any sorts of birds (except the few wingless species) are at all likely to become extinct; certainly they will not become so by being shot. Let the croakers think of the Pewit, whose eggs are so extensively gathered, both at home and abroad, and also of the numbers destroyed in harrowing and rolling the ground when they are sitting and laying; yet still the species is, perhaps, as numerous as ever. C. M. A.

In THE FIELD of November 17th, 1866, Mr. Edward Newman wrote that it had been very agreeable to observe the probability of this bird's breeding in Britain discussed in THE FIELD. He gave all the notices of its occurrence from 1789 to 1866, in 23 of which years it had been recorded. He added, he thought there was no sufficient reason to suppose the bird had ever bred here: had that been the case, July would have been the month to find the young (none, either old or young, were recorded in that month), and that he must regard the Little Bittern as one of those vernal and autumnal migrants that occasionally or accidentally visit Great Britain in the course of their migration.

This letter from such an authority should caution people from jumping at the conclusion that birds met with on migration late in going North, or early in returning, had bred in England, unless positive proof has been shown, and that proof is capable of being established by those able to give it, and at the same time thoroughly to be relied on.

The notices often seen of birds having occurred I may add are not unfrequently incorrect; far fewer of us know many species well enough to describe them than is thought of. I do not believe there is any one in England who is acquainted with all the birds *recorded* as English in the various states of plumage they assume at the different seasons and ages. This may be considered a wild assertion, but I believe it is no less true notwithstanding.

www.ingramcontent.com/pod-product-compliance
Lightning Source LLC
Chambersburg PA
CBHW030355170426
43202CB00010B/1385